Going With The Flow

Small Scale Water Power

Billy Langley

Dan Curtis

Centre for
Alternative
Technology
Publications

Going with the Flow
Billy Langley, Dan Curtis
© May 2004
all rights reserved

The Centre for Alternative Technology
Machynlleth, Powys Wales SY20 9AZ
Tel. 01654 702400 • **Fax**. 01654 702782
email: info@cat.org.uk • **Website:** www.cat.co.uk

Illustrations Hugh Moore and Graham Preston
Cover Photography Mansel Davies
Edited by Dave Thorpe and Stokely Webster

ISBN 1 898049 18 1
1 2 3 4 5 6 7 8 9 10

Printed in Great Britain by Biddles Ltd. (01553 764728)
on paper obtained from sustainable sources.

Billy Langley and Dan Curtis have asserted their moral rights to be identified
as the authors of this work in accordance with the Copyright, Designs and Patents Act 2004.

Published by CAT Publications, CAT Charity Ltd. Registered charity no. 265239.

Foreword

The UK Government has set a target for the electricity industry of providing 15% of the country's electrical energy from renewable energy by 2015. This ambitious aim will only be achieved if all those sources of renewable energy which can be used economically are developed. It will require the building and commissioning of about 600MW of capacity each year for the next 11 years.

By the nature of our climate and topography, hydropower can only be expected to play a minor, though not insignificant role. If even 25% of the 30,000 water mills which once operated in the country were each to generate 10kW, they would provide 75MW, and Scotland has at least 300MW of small-scale hydro undeveloped.

This book, *Going with the Flow* is a part historical, part technical account of what has been done and what can be done with small-scale hydropower plants. It is a primer for the layman who would like to be involved – perhaps to build his own small plant – and it takes the reader step-by-step through the process from the first thoughts to the technical, legal and economic aspects of implementing them. It will form a valuable introduction to anyone who wants to contribute, in however small a way, to achieving that target.

E M Wilson
Chairman 1995 to 2000
British Hydropower Association

Preface

Water power provides around 20% of the world's electricity, indeed for some 30 countries it is the main source of electricity. The great majority of this power is produced by huge multi-megawatt, sometimes even gigawatt-sized schemes. An increasing minority, however, is being produced by far smaller schemes, sometimes of just a few kilowatts in size...

Micro-hydro defines small electricity generating water power schemes with outputs ranging from as little as 0.5kW up to around 100kW. Compared with other methods of electricity generation, it is perhaps the most benign; pollution is not created, it is powered by a renewable resource. What is more, micro-hydro schemes tend to last—a well-designed system will provide a consistent and reliable mains-quality electricity supply for well over twenty years.

This book is written both for people who are considering the installation of such a scheme and for those with a general interest in the subject. It is intended to be clear and concise, avoiding unnecessary technical jargon, and accessible to all. Where unfamiliar terminology is encountered, the reader may take refuge in the comprehensive glossary provided towards the rear of the book.

Going with the Flow bridges the gap between small information pamphlets and full technical design manuals. To provide the reader with a basic grounding, much of the process of writing this book has involved summarizing the methods discussed in the design manuals, and simplifying the concepts. There are two such texts to which this book is particularly indebted; these are Allen Inversin's *Micro-Hydro Power Sourcebook* (1986, NRECA International Foundation) and Adam Harvey and Andy Brown's *Micro-Hydro Power Design Manual* (1993, IT Publications).

Acknowledgements

Billy would also like to thank Dulas Ltd and Richard Ramsey; his wife Sarah for her love and support over the years and his Dad for giving him the hydro bug.
Dan would like to thank the people of CAT for their help and assistance, especially those of the Engineering, Information and Publications departments and Stokely Webster for her rigorous and patient editorship.

Contents

Chapter One
Introduction to Water Power

BACKGROUND

Hydro power is a proven and mature technology, harnessed world-wide for over two thousand years. A traditional method of grain processing throughout the world, it also played a major role in the modernization and industrial development of Europe and North America.

Our use of hydro power probably began with the 'Noria' **water-wheels** built by the ancient Egyptians to raise water for irrigation purposes. These were a crude form of the **undershot** water-wheel, a design in which the paddles, or blades, dip into a natural stream of fast-flowing water, and the pressure of the water against the

————— The Undershot Wheel—————

Fig 1.1

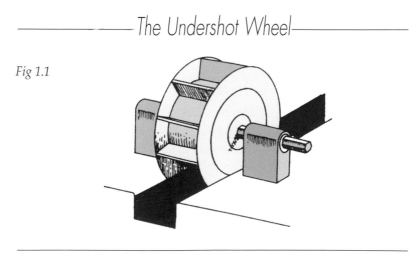

blades drives the wheel round on a **horizontal axis**. With the Noria, jars were connected to the rotating wheel such that some of the water was lifted to an elevated irrigation channel. The wheels were mainly constructed in wood, with a few of the moving parts made from metal or stone.

A few centuries later the first **watermills** appeared. A watermill is simply a water-wheel with a grinding apparatus attached to its shaft; the mechanical power of the water is harnessed to mill grain. The very earliest watermills were developed for milling corn in the Middle-East around 2000 years ago. These were unusual in that the wheel was spun on a **vertical axis**, with the water striking the blades from the side. This action can be seen in figure 1.2, the Ghata wheel, which is still used today throughout much of Nepal.

———————————*The Ghata Wheel*————————————

Fig 1.2

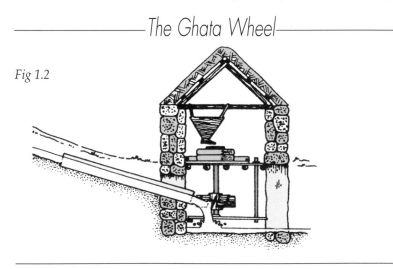

By the time of the Industrial Revolution, the use of water-wheels was widespread, the horizontal-axis **overshot** water-wheel being the norm. With the overshot design, water flows over the blades from above. The blades are built into the rim of the wheel such that they have closed sides, effectively becoming buckets. The weight of the water descending in the buckets turns the wheel, and this in turn produces power. Generally more efficient and less liable to flooding than the undershot wheel, the overshot wheel does have the limitation that it can only be used in situations where the height difference between the entering and departing water (the head) is at least equal to the diameter of the wheel, thus making it inappropriate for use in rivers with a gentle gradient.

Overshot Wheel

Aberdulais Falls (ETSU)

Later, the **breastshot** wheel was developed as a design compromise between the undershot and overshot wheel. With the breastshot, water enters at the level of the wheel's axle, thus requiring a lower head, while retaining a low susceptibility to flooding.

By the end of the nineteenth century, there were over thirty thousand watermills in Britain. The output from these mills ranged from a few kilowatts up to a few tens of kilowatts. Some of the larger versions powered machinery directly: ironworker's forges and power hammers or mine equipment such as water pumps, lifting tackle, and ventilation equipment. The use of iron components allowed for higher power to be handled, and several machines of 100kW or more were built.

——An Early Integrated Hydro-Electricity Generator——

Fig 1.3

The first water **turbine** was developed in France in 1832. A water turbine is distinct from a water-wheel in that it runs completely submerged; guide vanes direct the water onto the blades. The word 'turbine' is derived from the Latin 'turbo', meaning high speed, because it runs at a far higher speed than the traditional water-wheel. The smooth flow of the water through a turbine means that more of the energy of the water can be converted into useful mechanical power.

The development of the turbine, combined with the use of new materials, allowed the harnessing of water in different ways. Where there was a reasonable head available, the advent of cast iron piping allowed for water to be delivered to the turbine under pressure. This paved the way for increasingly powerful and compact, high-speed turbines which were not only cheaper and more efficient, but also far more flexible in design than the huge overshot water-wheels they were replacing. A little over half a century later, the potential for using turbines to spin generators and produce electricity was recognized.

The first electricity-generating turbines were installed around the turn of the century. Schemes ranged in size from 'large scale' (more than 5MW), through 'small scale' (hundreds of kW), to micro-hydro (less than 100kW). By the 1920s, the use of micro-hydro electric power in the UK had reached its peak: where there was the resource, independent hydro schemes were the main source of power for many towns and villages.

THE COMING OF THE GRID AND LEGISLATION

As the National Grid expanded into rural areas, the new 'mains' supply of electricity appeared far more attractive than the small hydro schemes: the Grid promised apparently limitless power, increased reliability, and low supply and connection costs. Later, with new nuclear power stations coming on-stream, it was widely thought that the price of electricity could only fall—according to the promoters of nuclear power it would become 'too cheap to meter'. Sadly, all but a handful of the independent micro-hydro schemes were abandoned and fell into disrepair.

The situation worsened as a result of new legislation which further deterred the small-scale generation of hydro power. The 1947 Nationalisation Act made it difficult, and often illegal, to generate and sell electricity privately. The Central Electricity Generating Board (CEGB) and local area electricity boards were established; apart from a very few exceptions, these held a monopoly over the entire electricity market. The situation was further compounded by the 1963 Water Resources Act which enabled the water boards in England and Wales to introduce charges for abstraction from any watercourse. As a result, very high annual abstraction rates were levied, with private commercial generators receiving huge bills from the water boards for the use of water in their schemes, even though the water was returned to the watercourse and established water rights existed. The Act effectively put a stop to all legal developments of small commercial water power schemes for a number of years.

In 1975, the National Association of Water Power Users (NAWPU) was established; but it was not until the 1989 Water Act, passed after intense lobbying by NAWPU and others, that the annual charges were waived for schemes of less than 5MW.

RENEWED INTEREST IN MICRO-HYDRO

The combination of unstable and rising energy prices in the early 1970s, together with more recent concern for the environment, has led to an increased interest in clean, safe, sustainable, and low-cost (over the long-term) renewable energy sources. Micro-hydro power is simple, reliable, and inconspicuous. Where there is the resource, it is widely recognized as perhaps the most attractive form of renewable energy. Micro-hydro is especially appropriate in more remote areas, where extending the National Grid poses practical difficulties.

Various technical developments, including advances in electronics, the introduction of lightweight, inexpensive plastic piping, and maintenance-free scheme components, have accompanied this renewal of interest, making cheaper, more reliable and flexible systems available.

Case Study: Chatsworth House, Derbyshire

Chatsworth House in Derbyshire boasts a spectacular array of water features in its gardens, including the Emperor Fountain—the highest fountain in Europe to be driven by gravity alone. Work began on the fountain in 1843, when the sixth Duke of Devonshire employed his head gardener, Joseph Paxton, to create a fountain that could rival that of Czar Nicholas, Emperor of Russia, due to visit the following year. Paxton set to work and in a little over six months had created a reservoir, the Emperor Lake, which provided sufficient pressure for the fountain to play to at least 75m. The lake occupied 3.5 hectares at an average depth of 2m, some 100m above the house; it was connected to the fountain by means of an 800m, 43cm internal-diameter pipeline.

Almost fifty years later, in 1890, it was decided to make additional use of the works of the fountain and generate electricity. The pipeline was extended to provide a head of 120m and diverted to a turbine house hidden to the side of the fountain. Three 'Gilkes Vortex' turbines were commissioned in 1893, and between them they generated just under 90kW—sufficient to meet the electrical needs of the house for over forty years. However, in 1936 the turbines were phased out in favour of connection to the National Grid.

Over half a century later, the Estate began investigating the possibility of recommissioning the turbines in order to reduce electricity costs. In 1988, work began on the installation of a new system, consisting of one single-jet 'Gilkes Turgo' turbine capable of producing up to 125kW of power. It was decided that the original pipeline would be used to supply the water in the first instance, with the option of laying an additional pipeline at a later date (an additional pipeline would double the capacity, and the turbine would become a two-jet unit capable of producing 250kW). By the end of the year, Chatsworth was once again powered by water.

——*Fountain and Turbine at Chatsworth House*——

Ian Fraser-Martin

ENVIRONMENTAL ASPECTS

There has been some publicity in recent years about large hydro power installations overseas that have proved environmentally disastrous. Problems have included the large-scale loss of habitats—both human and natural—as well as such hydrological effects as changes to water flows and groundwater levels, and the silting up of reservoirs. However, most of these problems only apply to poorly planned, very large-scale schemes. Carefully planned hydro power installations—especially those on a small-scale—should have minimal environmental impact. Indeed, any problems are likely to be far outweighed by the environmental benefits; specifically, generation of electricity by hydro power helps prevent climate change: it displaces the waste and pollution generated by fossil fuels or nuclear power.

On average, in any given year, a typical 10kW hydro-electric installation in the UK will prevent the consumption of the equivalent of 21 tonnes of oil and 36.5 tonnes of oxygen. It will also prevent the release of 70 tonnes of carbon dioxide, thereby helping to limit the effects of climate change, and one tonne of sulphur dioxide, thereby helping to prevent acid rain.

The energy used in the manufacture and installation of all components of a typical hydro-scheme, known as embodied energy, is generally matched by that generated by the scheme within 5–9 months of commissioning.

PROS AND CONS OF MICRO-HYDRO POWER

Micro-hydro power is an effective form of renewable energy… most of the time. Before embarking on a project, it is important to understand the limitations as well as the potentials of that system. Some of the important pros and cons of micro-hydro schemes are described below:

Pros

- Compared with other small-scale renewable energy systems, power is produced at a fairly constant rate. The power is available at any time and there is rarely any need for storage batteries.
- The technology is easily adaptable for manufacture and use in developing countries and remote areas.

- No primary fuels are required and maintenance costs are low.
- The technology is simple and robust, leading to lifespans of over 20 years without requiring significant further investment.
- In many cases, overall costs can undercut all alternative electricity sources—including mains supply.
- The principle problems faced by large hydro schemes, such as reservoir silting and the resettlement of communities, are not an issue. Micro-hydro power schemes tend not to require reservoirs; where they do, they tend not to be very large.
- Schemes emit no carbon dioxide or hazardous by-products, and little noise or waste heat.

Cons
- It is a site-specific technology, and suitable sites must be fairly close to locations where the power is required.
- On small streams, available energy is limited and cannot be expanded if demand grows.
- Output can be weather-dependent. Unless a small reservoir is incorporated into the scheme, power output may be limited during the summer months.
- The capital cost is high compared to a diesel generator, although this is offset by much lower running costs.
- There is a low-level environmental impact on the water course. Some civil works will be required, and over the section of the stream or river from which the water is diverted there will be less water available for other users.

COMMON MISCONCEPTIONS
There are some common misconceptions about hydro schemes which need to be dispelled before going much further.

A small amount of water cannot generate much power.
Fact: The power available depends entirely on two factors: how much water is flowing (known as flow), and the vertical distance over which it falls (known as head). A small amount of water can produce plenty of power if it falls far enough.

The amount of power depends on the speed of the water turbine.
Fact: Power is related to the product of head and flow. Different types of water turbines extract the potential energy of water in

different ways, and the speed of the water through the turbine depends on the scheme's head and the turbine's internal configuration. For a given turbine, there is an optimum operating speed for maximum efficiency.

A large amount of moving water will generate a lot of power, even at no head.
Fact: If there is no fall available, a large amount of moving water will not generally permit the generation of power. There must always be a head involved to provide potential energy. A large, slow river has momentum, but little other energy. River current turbines can be used in fast rivers (with a flow of more than 1.5m/s), but these are limited in application.

Micro-hydro schemes require a large reservoir, inundating many square miles of land, with all the associated ecological and social problems.
Fact: Most micro-hydro schemes do not need a reservoir since they generally operate as run-of-river, that is, water diverted from the flow is returned downstream.

Micro-hydro schemes kill fish and upset the ecology of the stream.
Fact: With careful design, the environmental impact can be minimized whilst only slightly compromising energy output. Environmental regulations require that systems are designed to leave a level of water in the stream adequate to sustain the life within it. Fine screens prevent fish entering the system and special 'fish ladders' can be installed to enable the fish to bypass any weirs.

Water turbines are noisy.
Fact: A well designed system can be virtually inaudible from outside the turbine house.

Hydro schemes are a blot on the landscape.
Fact: Intakes (civil structure in the river where the water is taken) can be stone-faced and discreet, pipelines and transmission cables can be buried, and turbine houses can be designed according to local character.

Water is consumed by the system.
Fact: All the water is returned to the stream, although there will be a stretch, between the intake and the turbine, with reduced flow. Only the potential energy of the water is extracted.

The water is of worse quality when it is returned to the stream.
Fact: The water is not contaminated in any way. In fact a hydro scheme often improves the oxygenation of the stream, and a weir can create a pond habitat for wildlife.

Micro-hydro schemes are unreliable.
Fact: With the development of maintenance-free water intakes and solid-state electrical equipment, a modern hydro scheme in a remote area is often more reliable than the mains.

The electricity generated is not of good quality.
Fact: Through the use of the latest alternators and electronic control equipment, a modern hydro-electric scheme produces electricity of exactly the same quality as the mains.

A house needs 13kW of power to function.
Fact: Over a day, a household's average electricity consumption works out at less than 0.5kW. Through simple load management, it is quite possible to reduce the peak demand to less than 3kW.

The scheme must run all year round to be worthwhile.
Fact: Income or savings are present regardless of how many months a year a scheme runs. In any event, the greatest demand tends to occur during the wettest months, when the scheme is most likely to be running to its full capacity.

Why bother with hydro power when all the country has mains?
Fact: The economics are undoubtedly better for those without a mains connection, but using an on-site hydro power resource can be worthwhile even if you are connected to the mains. Many people prefer renewable power to mains electricity and fossil fuels because it is non-polluting, and will be a good investment over the long term.

Hydro-power produces free electricity.

Fact: Hydro-power is capital intensive, requiring a sizeable investment. There are some ongoing maintenance costs too.

Chapter Two
Principles of Micro-Hydro Power

INTRODUCTION

Having looked at the history of the development of hydro power, and exposed some of the myths surrounding it, we move on to look at some basic principles. This chapter provides an overview of the physics required to calculate the quantity of energy stored in a body of water and, most importantly, how much of this energy can realistically be expected to be harnessed as electricity. The chapter also touches upon the basic principles of turbine operation and provides an introduction to the components of a typical micro-hydro scheme.

THE ENERGY IN WATER

Any body of water, anywhere, contains **energy**. Sometimes this is strikingly apparent—you can hear the energy of a waterfall, see the energy of churning white water rapids, or, if brave enough, jump in, and feel the energy as it tries to carry you away. Other times, with a sheltered and calm mountain lake, for instance, the energy is more subtle—but it is still there. The water in the lake is full of **potential energy**; should any of the water leave the lake and flow down the mountainside, its potential energy will gradually become transformed into **kinetic energy** (in addition, some of the water's energy will be lost—dissipated in the forms of sound, heat, and friction as it erodes a course through the rock).

A body of water has potential energy if it can fall to a lower level. Energy is measured in joules, and the quantity available at a particular site is dependent on a combination of elements: the quantity of water available, known as **flow**, and the vertical height through which the water can fall, known as the **head**.

With a micro-hydro scheme, the aim is to convert some of the potential energy of an elevated body of water into **mechanical** and then **electrical energy**. This is achieved by passing the water through a turbine.

Turbines
Turbines are discussed in detail later in the book. Here, the basic principles of operation are introduced.

Micro-hydro turbines convert the water's energy in one of two general ways, depending on the geographical location of the water source. In hilly or mountainous country where there is a high head available (over 20m), a compact, high speed **impulse turbine** is favoured. The water is conveyed to the turbine in a pipe; the potential energy of the water in the pipe is turned into kinetic energy as the water is projected out through one or more jets. This kinetic energy is then converted into rotational mechanical energy when the jets of water strike a series of buckets mounted on the outside rim of the turbine wheel (or, 'runner'). This mechanical energy is then transmitted via a shaft to the generator, which converts it into electrical energy.

The **reaction turbine** is often used in river valleys where the available head is low, but there is a compensatory higher flow. In contrast with the impulse turbine, there is no jet involved; an enclosed flow of water acts continuously against the turbine runner, producing a pressure drop and causing the runner to spin. These turbines tend to be more expensive than the impulse type since they are larger, more complex in shape, and more precisely machined.

In situations where there is no discernable head, a run-of-river propeller turbine, similar to a boat's propeller running in reverse, can be used. Here, the kinetic energy of the running water is harnessed through a reduction in its velocity. However, such schemes are not usually considered appropriate for micro-hydro applications.

POWER FROM THE WATER

When generating electricity, it is usual to talk in terms of power. Power is simply a measure of the rate at which energy is converted, it is usually measured in watts (joules per second).

Therefore, the potential power available from a body of water is a function both of head and of the quantity of water available per second (also known as the flow rate). It is described by the following equation:

$P = Q \times H \times \gamma$ where: P   is the power in watts (W),
Q   is the flow in litres per second (l/s),
H   is the head in metres (m),
γ is the specific weight of water (9.81 kN/m³).

Power – flow rate x head x specific weight of water
(W)   (l/s)   (m)   (kN/m³)

Alternatively, it may be more convenient to consider flow in cubic metres per second (m³/s) and power in kilowatts (kW). As 1m³ = 1000 litres and 1kW = 1000W, a direct equational substitution is possible—hence:

Power = flow rate x head x specific weight of water
(kW)   (m³/s)   (m)   (kN/m³)

From the equation, it can be seen that for every metre of head through which one cubic metre of water falls per second, there is 9.8kW (1 x 1 x 9.8 = 9.8) of power produced in the energy conversion. With a micro-hydro scheme, we are trying to harness the maximum possible portion of this power.

Example

One of the water turbines at CAT uses 20 l/s of water, with a head of 32m. From the above equation we can calculate the total amount of potential power in the water as it falls:

$$P = 20m^3/s \times 32m \times 9.8kN/m^3$$
$$= 6272W \text{ or } 6.3kW$$

This is the power in the water, however it does not reflect the amount of electrical power that we can actually expect to generate. In order to make a realistic calculation of that figure, it is necessary to assess the inherent inefficiencies involved in the conversion process.

Efficiency

It is inevitable that in the conversion of the potential power of the water to electrical power losses will occur. Taking these losses into account, the overall efficiency for a small scheme is usually considered to be around 50%—only half of the original potential power of the water can be expected to be converted to electrical power.

Fig 2.1 Efficiency of each element

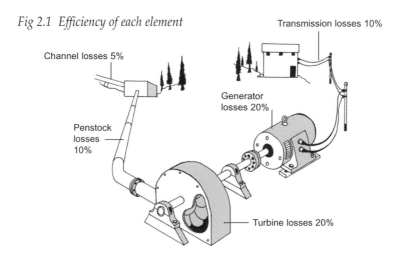

Transmission losses 10%

Channel losses 5%

Generator losses 20%

Penstock losses 10%

Turbine losses 20%

On this basis, the expected electrical power can be calculated using the following equation, sometimes known as the **Power Equation:**

$P = Q \times H \times \gamma \times e$ **where e is the factor used to account for (in)efficiency, that is, 0.5.**

Example

Assuming a 50% efficiency for the system at CAT, we can expect the following results:

P $= 20 \, l/s \times 32m \times 9.8 \, kN/m^3 \times 0.5$
     $= 3.1kW$ Electrical Power

(In fact, the efficiency factor is a little better than 0.5: 3.5kW is actually produced.)

The Pelton turbine at CAT

COMPONENTS OF A MICRO-HYDRO SCHEME

In order to generate electrical power, the water must first be directed to the turbine. The following diagram introduces the components that might be required in a typical micro-hydro power scheme.

————Elements of a Micro-Hydro Scheme————

Fig 2.2

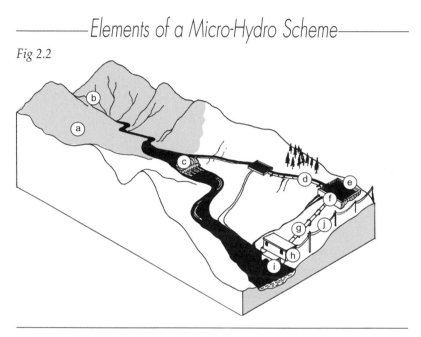

a) **Catchment area:** The area of land above the intake that provides the springs and rainwater run-off from which water is collected and passes into the stream.

b) **Watercourse:** The body of water—stream or river—from which the power is to be harnessed.

c) **Intake:** The point in the scheme where a controlled flow of water is diverted from the stream or river; often incorporating a diversion weir.

Water intake diversion weir: Most schemes considered here are 'run-of-the-river' so they do not require any water storage. The main function of the weir in such schemes is to seal the riverbed and stabilize the flow, allowing some of it to be diverted into the scheme.

Intake orifice: The orifice built into the intake and sized to allow the design flow to pass into the channel, but restrict the flow during floods.

d) **Channel:** Also known as the head race canal, a channel is often used to convey the water over the distance from the intake to the settling tank at the entry to the penstock. The channel is open and is usually lined in order to optimize the flow and minimize the risk of leakage.

e) **Settling tank:** A large tank at the end of the channel which slows the water down sufficiently to allow debris to settle out prior to the water's entry into the penstock. A flushing valve for clearing the accumulated gravel and silt is usually incorporated.

f) **Trash rack:** This is simply a screen, fitted across the entry to the penstock, which prevents leaves, debris and fish from entering the system. New, maintenance-free trash racks with very fine spacings are available, and these can obviate the need for a settling tank.

g) **Penstock:** The pipe used to convey the water to the turbine. A variety of different pipe materials may be used; however, all tend to be expensive, so it is always better to have as short (steep) a penstock run as possible—often necessitating the use of a channel.

h) **Powerhouse:** The building at the end of the penstock housing the electromechanical equipment—the turbine, generator and governor.
Turbine and generator: The hardware for converting the potential energy of the water into electricity. Known as the electromechanical equipment.
Governor: Not the boss, but a device for keeping the speed of the turbine and generator constant (if not connected to the grid) by controlling either the flow of water to the turbine, or the electrical load on the generator.

i) **Tailrace:** The pipe or channel to return water from the powerhouse back to the watercourse.

j) **Transmission cable:** Buried or overhead electric cable for transmitting the electricity from the generator to the load.

Chapter 3
Micro-Hydro Power
Practicalities

INTRODUCTION

Micro-hydro schemes may vary in size from a few hundred watts to several tens of kilowatts, depending on their application. Schemes may be characterized as domestic or commercial. Domestic schemes either supplement an existing mains electricity supply, or provide an autonomous, 'stand-alone', electricity supply. Commercial schemes produce electricity to sell.

This chapter looks at the practicalities of living with domestic micro-hydro schemes; it is especially relevant for those considering the installation of stand-alone systems where there is no mains electricity connection to back you up.

What makes a good micro-hydro scheme?

A good micro-hydro scheme is one that produces as much power as you need as cost-effectively and reliably as possible. How much power is needed will depend very much on what the power is expected to do.

With a commercial scheme the object is to sell the power. If it is to be sold through the National Grid then the demand for your power will, in effect, be limitless—whatever you can produce will be absorbed. However, because of, amongst other things, the costs involved in installing the equipment required to make sales to the Grid, such systems will generally need to produce at least 25kW before becoming viable.

With a domestic scheme designed to supplement the basic energy needs of a house with an existing mains electricity connection, the object may be to reduce bills and minimize environmental burden. However, because there is a 'Grid connection' to provide back-up, the quantity of power produced by the hydro scheme is not terribly important as any shortfalls will be met. What *is* important is that *some* of the house's power demand is being met.

With a stand-alone domestic scheme the situation is somewhat different. As there is no mains electricity supply to fall back on, the system must be capable of providing sufficient power to meet *all* your requirements. However, the more power you require, the bigger and more expensive the system becomes. Therefore, it is essential, before designing such a system, to ensure that your power requirements are kept to an absolute minimum. Fortunately, this can usually be achieved quite simply.

This chapter describes how to assess and manage your demand such that maximum utility can be derived from the minimum amount of power.

SUPPLY CONSIDERATIONS

Before considering power requirements, thought should be given to power availability. At a given site, the availability of energy and power, or the resource potential, will be dictated by the available head and the available flow.

Available head

The available head depends on the topography of the land over which the watercourse flows, and the extent of the water course that is available to be used (generally governed by land ownership boundaries). To achieve the maximum possible head, the intake and the powerhouse would be placed at the highest and lowest points, respectively, within the site's boundaries. However, economics usually dictate that only part of the watercourse can be made available for a scheme—preferably the steepest section with the greatest catchment area. If the resource is a steep mountain stream, it may be possible to obtain more head fairly easily. But there is a pay-off between head and flow: flow reduces towards the top of the catchment. Therefore, care should be taken to avoid getting too close to the source of the stream, as most of the catchment area will be lost.

The lengths of the penstock and channel (if used) are important economically: in order to minimize frictional losses, longer penstock runs require larger diameter pipe, therefore costs, being a function of both diameter and length, rise disproportionately. In order to minimize length, the average gradient of the penstock should be at least 1:20, ideally closer to 1:10, especially on small schemes.

Natural features, such as the crests of waterfalls, often make ideal intake sites.

Available flow
The average available flow depends on the catchment area size and run-off characteristic, the rainfall, and evapo-transpiration (water returned to the atmosphere by evaporation and plant transpiration). The flow rate at a given time of year will vary enormously, from flood to drought. The average flow rate or mean flow is defined as the total volume of water flowing down a river in a given period of time divided by the number of seconds in that period. Since flows are generally measured on a daily basis, this is called the average daily flow (ADF). Particular years have their own ADF. In any given stream or river, the ADF is generally exceeded for at least 25 to 35% of the year (the larger the catchment area, the greater the percentage). In some areas, annual flood flows can be as high as 100 times the ADF, whilst a drought flow might be only 100th of the ADF.

Assessment of resource potential is covered later in chapter 4.

DEMAND ASSESSMENT
The electrical demand in a typical household varies greatly over the course of a day, and throughout the year. Therefore it is important to draw up an overall profile of the expected demand. In order to calculate the figures, a table such as 3.1 may be used.

Table 3.1

Appliance	Power consumption (watts)	No.	Demand	Hrs/day	Total Wh/day
Cooker	5000 (max)	1	5000	1	5000
Fridge	100 (av)	1	100	24	2400
Toaster	1000	1	1000	0.25	250
Kettle	2000	1	2000	1	2000
Microwave	650	1	650	0.5	325
Television	100	2	200	1.5	300
Video	120	1	120	1	120
Stereo	50	2	100	2	200
Lights	60	10	600	4	2400
Vacuum cleaner	1200	1	1200	0.25	300
Washing machine	500	1	500	0.5	250
Computer	150	1	150	3	450
Printer	50	1	50	0.25	13
TOTAL			**11670**		**14008**

Maximum possible peak instantaneous demand:	11670W
Total energy required per day:	14008Wh
Average demand:	584W

For an average demand of less than 0.6kW, it would not make economic sense to install a scheme large enough to meet the occasional 11.7kW peaks. Instead, it will be necessary to manage the demand. A variety of techniques are available.

DEMAND MANAGEMENT
Averaged out over a year, the typical household in the UK consumes approximately 500W at any given time. However, this is subject to some considerable variation. During certain parts of the day, the demand may be down to almost zero—in the early hours of the morning for instance. At other times, such as on an early winter's evening, demand could quite conceivably rise to perhaps 20kW, especially if there is a heavy reliance on electricity for heating and cooking purposes.

To allow for these occasional peaks in demand, a standard domestic connection can provide 14–24kW. With a small hydro scheme, such a luxury is unlikely to be available: in order to get the most out of the site, demand must be managed and peak requirements reduced.

A typical domestic hydro system might have an output of somewhere between 2–6kW, and this will be more than adequate to provide sufficient power most of the time. However, by making a few adjustments, it should be possible to ensure that there is sufficient power all of the time. Electricity should be used for equipment that cannot be powered by any other means—TV, video, computers, hi-fi, lighting, etc.; alternative fuels should be used to provide energy for the other (coincidentally, most power-hungry) appliances—for example, cookers and heaters should be gas-fired or solid-fuel versions wherever possible. At a pinch, lights may even be gas- or paraffin-powered. Of the electrical appliances that are used, the following points may be considered:

• Energy efficiency is perhaps the most important concern. Compact fluorescent light bulbs use a fifth of the power of standard bulbs for a given light output. Most 'white goods' (fridges, washing machines, etc.) now carry 'energy labels' detailing their relative efficiencies, which help make it possible to buy the most appropriate units.

• A further point with white goods—especially fridges and washing machines—is that they use electric motors. Electric motors can use two to six times their rated power consumption on start up. A 100W refrigerator may need up to six times its rated power—0.6kW—to start. Fortunately, short bursts of power in excess of the rated power can usually be supplied from the inertia of the hydro system's rotating machinery.

• Low power versions of standard electrical appliances can also ease the load burden. Standard electric kettles consume around 2kW; however, low power 0.75–1kW models are available and take only slightly longer to boil. Slow-cookers consume a steady, low (less than 100W) amount of electricity over a long period, making them more appropriate than other forms of electrical cooking.

• Finally, good-housekeeping should ensure effective demand management. Obviously, lights should be switched off when not in use; but care should also be taken to ensure that when any high consumption appliances are used, they are not used at the same time—for example, a power drill should not be used while the washing machine is on, etc.

By simple load management, and use of some of the alternatives discussed, the peak instantaneous demand may be greatly reduced. Table 3.2 calculates the revised load.

TABLE 3.2

Appliance	Power Consumption (watts)	No.	Demand	Hours/day	Total Wh/day
Fridge	100 (av)	1	100	24	2400
Microwave	650	1	650	0.5	325
Television	100	2	200	1.5	300
Video	120	1	120	1	120
Stereo	50	2	100	2	200
Low energy lights	12	10	120	4	480
Vacuum cleaner	700	1	700	0.25	175
Washing machine	500	1	500	0.5	250
Computer	150	1	150	3	450
Printer	50	1	50	0.25	13
TOTAL			**2690**		**4713**

Maximum possible peak instantaneous demand:	2690W
Total energy required per day:	4713Wh
Average demand:	196W

Two simple measures have been taken:

• Appliances that consume less energy have been substituted. The electric cooker has been replaced with a gas (LPG or mains) version; the kettle replaced with a stovetop version; the toaster ditched in favour of the cooker's integrated grill; and a less energy-consumptive vacuum cleaner substituted for the standard model.

• Low energy, compact-fluorescent lightbulbs have been used.

As a result, the maximum possible demand is now just a quarter of the original figure. It would now be quite feasible to meet the electrical needs of the household with a 3kW hydro scheme.

POWER MANAGEMENT
Ballast ('dump') loads and heating

In the previous section we saw how the electrical power consumption of a typical household can be dramatically reduced through effective demand management. By adopting the measures described, the type of system required has been reduced to a quarter of its original size. However, glancing back at the figures, it becomes apparent that while a 3kW scheme will be necessary to cope with the peak demand it will also, on the basis of average demand, be providing a great surplus of power. Because the average demand is just 196W, the surplus would be approximately 2.8kW at any given time. Fortunately, this surplus need not go to waste. Indeed, such a scheme could be configured to not only meet the electrical demand of the house, but to also provide most of its space and water heating needs.

With most micro-hydro schemes, an electronic governor (known as an Electronic Load Controller) is used to manage the supply. Any spare electricity is diverted to either space or water heating as a spin-off. By assessing the average demand, the amount of energy which will be 'spare' can be calculated. A simple comparison of these figures with those for current or predicted heating fuel uses will indicate the impact this spare energy will have on space and water heating bills. The greatest demand for heating tends to correspond with wet times of the year, i.e. winter, when there is plenty of water available, making hydro power a good renewable source of heating.

In the above example it can be seen that were a 3kW generator to be used, 2.8kW of electricity would be available to be diverted to heating at any given time.

Battery charging systems

The majority of micro-hydro schemes do not use batteries, owing to the fact that the resource is continuous and there is no need to store the power. However, there are exceptions, especially with very small schemes.

Direct ac systems (i.e. systems with no battery store) with an output of less than 2kW can have difficulties in meeting the instantaneous start-up demands of such things as washing machines, for

——— Schematic of a Typical Domestic System ———

Fig 3.1

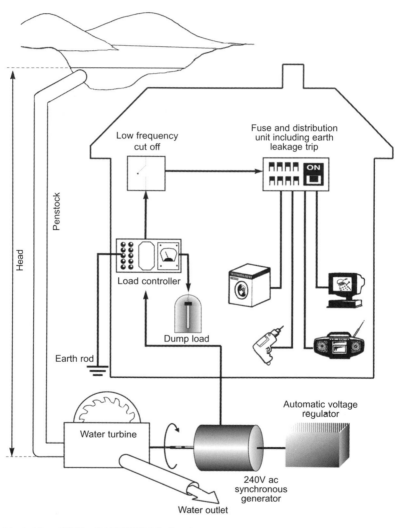

Adapted from *Off The Grid* (CAT Publications)

Case Study: 6kW stand-alone system

Here, the users had previously relied mains electricity for power, but wanted to become more self-sufficient in terms of energy use, and to reduce the environmental impact of their lifestyles. A medium-sized stream on the property was tapped via a low diversion weir with a maintenance-free intake screen. The scheme consists of a 200m long, 200mm diameter penstock pipe, feeding a four jet, 250mm diameter Pelton wheel connected by a belt drive to a 1500rpm synchronous alternator. It provides 1–6kW of continuous power—adequate to meet the electrical needs throughout the year and heating demand for most of the year (the remaining heating demand is met with fuel). The system operates on a head of 23m and a flow of 10–48 l/s, depending on the availability of the water. The total cost of the scheme was £18,000 installed. The annual savings on electricity and heating are on the order of £2000. The simple payback of around ten years illustrates the long-term investment nature of domestic schemes. However, it has a high feel-good factor, and the major components of the scheme will last in excess of thirty years.

Case study: Small 700W system

Here, the essential power requirements were originally met by a small petrol generator, and paraffin or candles supplied lighting. A small mountain stream on the property was tapped. The scheme consists of a 63mm flexible pipe for the penstock, a home-made 100mm diameter Pelton wheel turbine, a 3000rpm induction generator, and an electronic governor. This provides up to 700W continuous power, which meets the electrical and water heating requirements for most of the year. The system operates on a head of 60m and a flow of 1–2.5 l/s, depending on the weather conditions. The scheme cost around £3000, a fraction of the price of a Grid connection, whilst previously around £600 per year was spent on petrol, paraffin, and candles.

——————————Two Pico-Turgo Turbines——————————

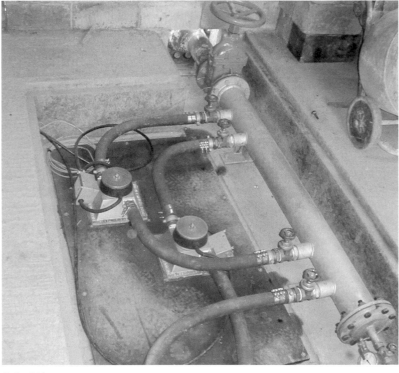

Dulas Ltd

instance. Having said this, some people are happy with a direct ac system of less than 600W capacity, using it to power lighting, audio visual, and other small loads, and accepting its limitations on larger loads. Another option is to run a diesel generator for a few hours each day or week in order to accommodate the heavy loads.

However, if the stream cannot meet the peak demand, yet your average demand is low, a small hydro installation with battery storage might be the best practical and economical solution. Such a system allows for the storage of power until it is needed. Generally the most cost effective way of doing this is to use purpose-made, lead acid batteries which store low voltage (12, 24, or 48V) dc electricity. Special 'deep-cycle' lead acid batteries are used (these differ in composition from car batteries, which are designed for heavy current draw off for very short, intermittent lengths of time).

—Schematic of a Typical Battery Charging System—

Fig 3.2

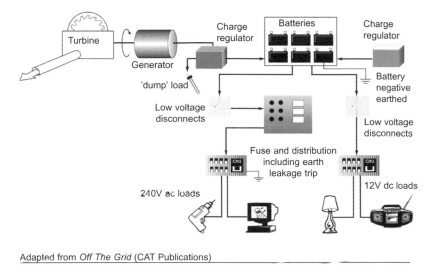

Adapted from *Off The Grid* (CAT Publications)

Many standard, low voltage electrical appliances are now available that can be powered directly from the dc electricity stored by batteries. If 240V ac equipment is to be used, an electronic device called an inverter can convert the low voltage dc stored in the batteries to mains quality ac, but some power is lost in the process. Using dc straight from the batteries avoids the energy losses associated with the inverter. However, thick cable is needed in the transmission of low voltage dc to limit losses. (Power is lost due to resistance in the cable; resistance increases with cable length and reduces as diameter increases.)

Solar and wind power can combine well with a battery charging hydro scheme. The electronics are relatively simple in either case, and with solar photovoltaics there is the bonus that they tend to produce most power during dry periods when a hydro scheme is most likely to be out of service.

For more details on battery-based systems the reader may wish to consult the CAT publication *Off the Grid*.

Case study: Battery charging system for a remote valve operation

In this case, the owner is a water company who required power for some control and monitoring gear on one of its pipelines in a remote location. The average power requirement is very low, at about 100W, but with peak demand reaching 1.2kW during valve operation.

A small 100mm diameter Pelton runner is used to drive a 24V generator and provide continuous charging for a battery store with sufficient capacity to power the equipment during valve operation. The water is provided by a 38mm diameter tapping from the pipeline, and the system operates at an equivalent head of 50–80m depending on the pressure in the pipeline.

The equipment cost around £5000 installed. This compares well with the alternatives of either a Grid connection, at £15,000, or the cost of periodic battery charging, at around £3000 per year.

SUMMARY

Having given consideration to the practicalities of micro-hydro power it should be possible to get some idea of how best to develop an appropriate scheme. To summarize, this will be dictated by the following factors.

Peak demand

This determines the maximum power output of your scheme. Ideally the scheme will match your peak demand. In fact, this is rarely the case—it generally makes sense to limit the peak demand and install a smaller scheme.

Average demand

The average of your total energy requirement (i.e. both electricity and heat) often gives an indication of the appropriate size for a scheme. The size should be sufficient to meet, at the very least, the average electrical demand. However, as described earlier, this will not meet demand peaks. Ideally the scheme will provide sufficient power to meet modest peaks in demand. The difference between this and the average demand can be dumped into heating loads.

Budget
Obviously a smaller scheme costs less money, so this has a bearing on the size of system you install. However, there are many fixed costs, so costs are not directly proportional to size. Unlike wind and solar systems, which are largely modular, with an output proportional to cost, a significant proportion of the cost in a hydro scheme is made up from the design, consultation, and installation. These costs are fairly fixed over a range of power outputs. Hence it is often worth installing a scheme larger than the minimum requirement—a 6kW scheme can cost little more than a 3.5kW scheme, and the additional power can be used for heating. Other costs are often standard too, such as the turbine, alternator, and governor. However, you should also consider the resource and the part-load efficiency of the equipment. If your stream is relatively small, there is little point in installing a turbine that operates at full power for less than a third of the year. Instead it would be better to go for a smaller scheme, operating on perhaps half the flow of the bigger one, but giving better performance during dry spells.

Available supply
Finally, the scheme must operate within the limitations of the resource. There is no point in installing a stand-alone scheme that only provides full output for 10% of the year. As a rule, a stand-alone scheme should provide full power for at least 50% of the time. If the available supply is very good, you could consider installing a scheme that is bigger than your minimum requirement. Often on small domestic schemes, the supply is the limiting factor, and the size is determined by this.

The following chapter gives a comprehensive description of how to evaluate the resource potential of a given site.

Chapter 4
Site Evaluation

By this stage you will have some idea as to whether or not it is worthwhile looking further into the possibilities of developing a scheme. This chapter explains the various methods used to measure the head and flow, and therefore assess the size of the available resource.

INTRODUCTION
The evaluation of a potential site is generally done in two stages: the desktop study and the site survey.

Before spending any money, it makes sense to carry out a brief desktop study of the site to assess its likely performance. This is undertaken using large-scale (preferably 1:25,000) maps of the scheme's location to assess the head and catchment area, and hydrological data to generate an approximation of the annual flow pattern.

The study should give some idea of the approximate resource potential and estimates of how well the scheme should perform throughout the year.

On the basis of the data gathered in the desktop study, a site survey is carried out: on-site measurements are made of the available head, and flow data is taken in order to check the hydrological modelling. These measurements will allow for a precise layout of the scheme to be put together. The techniques required for the site survey are described in this chapter; chapters 5–8 describe the detailed design, specification, economic analysis, and legal matters you will need to consider in order to complete a full feasibility study.

Consultants may be required to carry out some or all of this work, depending on the scale of the project, and produce a feasibility report with budget costings.

THE DESK-TOP STUDY

Large-scale maps are excellent tools for making a quick initial assessment of a medium to high head site. 1:25,000 scale Ordinance Survey maps with height contours at every 5 or 10m may be used. The most economic location for the penstock run will be steep, so look for an area on the map with contours that are close together. Be aware that much of the ground detail is only accurate to, at best, 5m, hence a map, as described, will not be suitable for assessing low head (less than 20m) sites.

The first stage is to produce a preliminary site layout.

Site layout

Layout is largely governed by the topography of the site. Some sites have obvious intake and turbine positions—for example, the top and bottom of a waterfall. When this is the case, the head is measured between these two positions. At other sites, however, it is not always so obvious, and an accurate survey of the steepest sections should be carried out.

The most logical location for the powerhouse is usually where there is an obvious reduction in the gradient of the penstock. Ideally the powerhouse should also be located close to where the power is needed—however, penstock pipe is more expensive than electric cable and there is usually little point in extending the pipeline on a near horizontal section just to bring the powerhouse closer to the load.

When locating the powerhouse and designing the intake, make allowance for the possibility of floods. The powerhouse should be sited clear of any flood areas. With the intake, the damage risk may be less of a cause for concern—in streams prone to severe flooding, it sometimes makes sense to install a low-cost intake that can be cheaply repaired after damage.

When there is no obvious site for the intake or turbine, a number of positions may be considered in detail. The pros and cons of each position are discussed in chapter 5.

Available head—quick approximation
Firstly a large-scale map of the area under consideration should be acquired and, if necessary, the relevant section enlarged. The total available head is the difference in altitude between the intake and turbine positions, measured off the map contours.

Available flow—quick approximation
The annual average daily flow (ADF) can be approximated using the following equation:

$$\text{ADF} = \frac{(A \times R)}{s}$$

Where: A is catchment area (m^2),
R is annual run-off depth (m),
s is number of seconds in the year (32×10^6)

The catchment area can be determined from the contour lines and run-off directions on the large scale map. The run-off is the amount of water, expressed as a depth, that is expected to enter the stream over the course of the year. It is equal to the annual level of rainfall over the catchment minus the portion of that rainfall that evaporates before it can get to the stream.

The average annual rainfall can be estimated from rainfall (or isohyetal) maps. Such maps simply have rainfall contour lines as opposed to height contour lines, and are usually available through the Meteorological Office, or, in the UK, the Institute of Hydrology, who charge less for the information. They tend to be small scale and are therefore only useful for making initial approximations.

The quantity of water either evaporating or transpired by plants and trees will depend on the weather and the soil conditions. It should be possible to obtain a 'rainfall vs. run-off chart', again from the Meteorological Office, for the area being considered. From this, a fair estimation of the likely annual run-off can be made on the basis of the annual rainfall level.

A typical domestic scheme might be designed to run at full power 40–50% of the year, with less flow available at other times. Design flow must be considered to be less than the ADF since much of the catchment's flow cannot be utilized during high flows and floods. Depending on the characteristics of the catchment, the design flow should typically be 50–75% of the ADF. On the other hand, for optimum returns, a typical commercial Grid-connected scheme will have a design flow approaching the ADF.

Example

A brief example of a desktop study is given for a small household scheme. With reference to the enlarged section of 1:25,000 map depicted in figure 4.1, it has been decided to install the intake at point A, with a channel running to point B, from where the penstock will run to the turbine at point C.

Fig 4.1

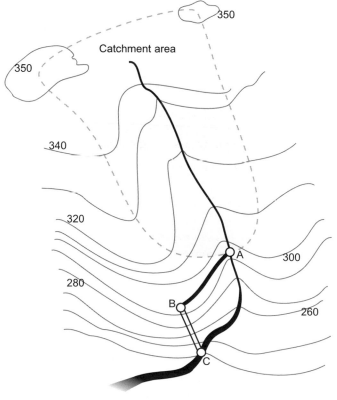

Estimating the head
Very simply, the head is the difference in altitude between the forebay
(B) and turbine (C), approximately 40m (±5m). Note that the penstock
run has been minimized through the use of the channel—its gradient
is steeper than 1:8.

Estimating the potential flow
The catchment boundary can be determined from the contour lines
and run-off directions on the map. In this case, the catchment area has
been drawn in and measures approximately 1km² (= 1 x 10⁶ m²).
From an isohyetal map of the area, the annual rainfall is found to be
2m, and, from a rainfall versus run-off chart, the evapo-transpiration
to be 0.5m. Hence run-off equals 1.5m.
With figures for the net rainfall and the area of the catchment, the ADF
can be calculated:

$$\text{ADF} = \frac{(A \times R)}{s} \text{ m}^3/\text{s} - \frac{1 \times 10^6 \times (2.0\text{-}0.5)}{32 \times 10^6} = \frac{1.5}{32} = 0.047 \text{m}^3/\text{s}$$

$$= 47 \text{ l/s}$$

Because this study is for a household scheme which would require the
turbine to run at full power for most of the time, the design flow will
be less than the ADF—say 65%.

Hence: **Design flow** =   ADF x 0.65   **= 30 l/s.**

Likely power output
Using the power equation described in chapter 2, an initial estimate of
the likely power output of the scheme can be calculated using the
figures for the head and flow (assuming an efficiency of 50%).

Likely power	=	$Q \times H \times \gamma \times e$	Q	design flow (l/s),
output			H	head (m),
			γ	specific weight of water (9.81 kN/m³).
			e	efficiency
	=	$30 \times 40 \times 9.8 \times 0.5$		
	=	5880W	=	**5.9kW.**

This should be plenty for a domestic system.

SITE SURVEY

Once basic feasibility has been proved through the desktop study,
a site survey will be required to assess the potential in further detail
and enable detailed design. Since the total available power of a site
depends entirely on the head and available flow rate, these two
factors will need to be measured more precisely than the initial
assessment permits.

Once a site layout has been determined, the gross available head between the intake and turbine position is measured.

When carrying out a survey, it is useful to be equipped with a notebook, as well as some numbered marking pegs to indicate the measuring points for future reference. Note down the positions of measurements, using either peg number or natural features.

MEASURING HEAD

The head can be assessed through a number of methods that use a variety of different equipment.

For optical methods:
- a builder's dumpy level and staff
- a spirit-level
- a theodolite

For pressure methods:
- a water-filled hose and staff
- a water-filled hose and pressure gauge
- an altimeter

All of these methods can be valid; the most appropriate will depend on site conditions, equipment, time available, and budget constraints. The equipment and its use is described below.

Optical methods

Perhaps the most practical way of taking accurate head measurements is to use a **dumpy-level and measuring staff**. The dumpy-level is a popular optical surveying device with a telescopic eyepiece and cross hairs. To ensure that it is level, it is mounted on a tripod with an adjustable base.

The site's head is the difference in vertical height between the proposed intake and discharge points. It is determined by measuring the change in height in convenient stages, as shown in figure 4.2.

The drop at each stage is tallied up to give the overall fall. It always pays to check measurements at least once, especially on low head sites. To do this, return to the starting position from the final point, that is, measure the site in reverse to complete a measuring loop. The overall head around the loop should come to zero, or thereabouts. This method is a favoured approach: it is potentially

———————Measuring the Head———————

Fig 4.2

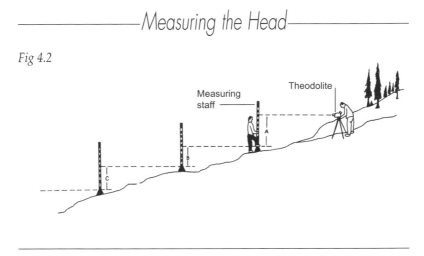

very accurate (±1%), it permits self-checking when completing the measuring loop, and the equipment is readily available. Difficulties arise when the site is heavily wooded, since this method depends on line-of-sight. Also, this method can be very time consuming if measuring a head over 50m.

In some cases it may be possible to use a **spirit level** as an alternative to the dumpy level, by mounting it on a plank, holding it level and sighting onto a staff. Accuracy will be limited because of inevitable errors in the holding of the level and sighting—using a plank and level will have an accuracy of 5–20%, depending on the number of measurements required.

At the other end of the spectrum is the infrared **theodolite**. Probably a bit over-the-top for most small schemes, such a device can provide a very accurate map, not only of the height contours, but also of the relative positions of the site's major features.

The device measures exact positions from a datum (fixed at the survey), by beaming an infrared light at a staff, onto which is mounted a special prism mirror which reflects the beam back. On the basis of the reflected beam, the theodolite is then able to calculate the relative position of the staff.

In use, the operator sets up the instrument at a good vantage point, while an assistant holds the staff at the major points of interest. The operator then takes readings of the staff's positions. Often, the device is used in conjunction with an automatic data recorder which downloads all the information onto a computer and automatically draws up a plan. This is how most landscape surveys are carried out, with the option to enter features such as trees and outcrops or roads. It is a very accurate and expensive instrument which is best used by a trained person.

Pressure methods
Of the pressure methods, the use of a **water-filled hose and measuring staff** is possibly the most basic. Using the water-filled hose is cheap and easy, but rather long-winded. The two hose pipe ends are level when you can fill the pipe with water and none spills out either end. You then drop a plumb-line from the top of the hose to the ground and measure this height on the measuring staff. Light and easy to use, this method's accuracy ±5%.

——Water-filled Hose and Measuring Staff——

Fig 4.3

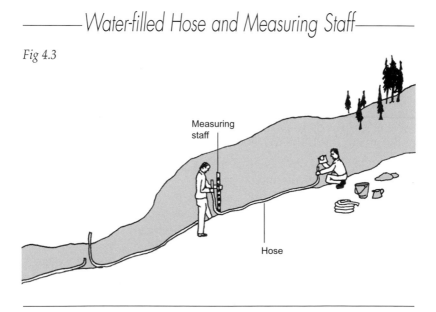

A slightly more advanced alternative is to attach a **calibrated pressure gauge** to the bottom of the length of hose. The gauge must be calibrated by checking it against a reference height, and the pipe must be free from entrapped air bubbles. Height can be calculated according to the reading on the pressure gauge. The advantage of this method is that only one reading needs be taken, and the hose can also act as a measuring unit for penstock length.

head (m)   = pressure in bar x 10, or
            = pressure in psi x 0.6895

———— Water-filled Hose and Pressure Gauge ————

Fig 4.4

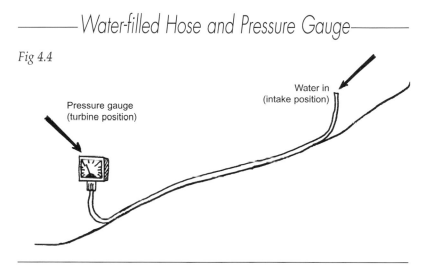

Easiest of all is the **altimeter**, which is useful for assessing medium to high head (50–300m) sites quickly. It works by measuring the change in air pressure that occurs with height. A reasonably priced (£200–300) sports instrument can be used. Take readings of head at the proposed intake and powerhouse sites several times. This method is weather and temperature dependant, so ideally it should be performed during stable conditions. Ideally, to account for atmospheric pressure changes between the measurements, a second altimeter should be used to plot the change in base pressure over time at a fixed altitude. So long as the timings of both instruments' readings are recorded, any atmospheric pressure 'drift' can be allowed for.

Accuracy can be as good as ±2%, but there is the chance of gross errors of ±30%. It not unknown for cheap altimeters to record a temporary gain in height as the operator is descending the hill! For accurate measuring, a very expensive surveyor's altimeter is required, which must be pre-calibrated.

Summary

Since the power output and design of the scheme components depend on the head, it is important that the measurements are accurate to within at least ±5%. Therefore, whatever method you use, re-check your results.

ESTABLISHING AVAILABLE FLOW

The volume of water available per second, or flow rate, is the other resource that determines a scheme's power output. The flow in a natural watercourse varies massively throughout the year, depending on the weather. You often find that the site will struggle over summer to supply the bare minimum required, whereas over winter there will be a surplus which may coincide with an increased energy demand. On the other hand, there can be long, dry spells in winter, with deep frosts and clear skies.

Overall energy production is not the only figure that concerns the micro-hydro user. It is important to know how the flow varies in order to predict when the most power will be available. The science of predicting flow is hydrology, and a hydrological study is the best way to assess a site's potential.

———————— The Hydrological Cycle ————————

Fig 4.5

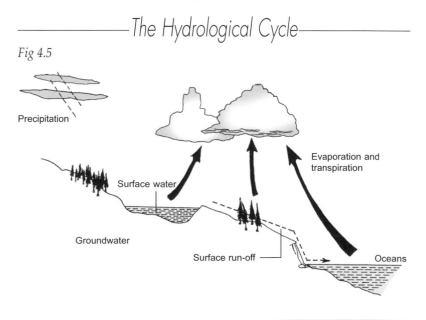

Flow is the hardest factor to predict, since it is a complex process based on the hydrological cycle, which describes the movement of the Earth's waters from evaporation to condensation in relation to the land. Driven by the energy of the sun, the Earth's water cycles continuously between clouds, surface water, ground-water and the oceans.

This section looks at the four stages involved in establishing the flow:

- getting to grips with hydrological models
- using the models to derive flow data for your site
- taking actual on-site measurements of flow
- accounting for likely seasonal variations

The hydrological models
The first stage involved in estimating flow data for your site will be to compile all the data already available in the public domain. In most countries there will be a body responsible for collecting data on flows in watercourses; the appropriate authority should be able to present data from a local gauging station. In England and Wales, the Environmental Agency maintains gauging stations on many rivers, and this data can be obtained at a small cost.

There are several useful ways of expressing the hydrological data and describing how the flow in a watercourse varies throughout the year; it will generally be provided in tabular form or represented graphically, as either an annual hydrograph, or a flow duration curve.

Annual hydrograph
An annual hydrograph is simply a graphical representation of variations in flow over time for a given watercourse, over a period of one full year. The graph will only be valid for the particular stream or river in question.

The hydrograph gives an idea of what flow levels can be expected, and when. What is immediately apparent is how the flow rate varies massively throughout the year. It is also interesting to observe that the flow rate tends to rise quickly but drop off more slowly.

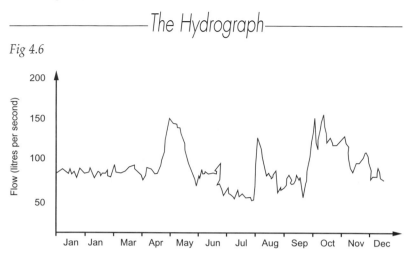

The Hydrograph

Fig 4.6

Flow duration curve

A flow duration curve shows the relative duration of the range of flow experienced throughout the year. This is generally expressed as flow rate versus the percentage of time during which it is exceeded, as shown in figure 4.7.

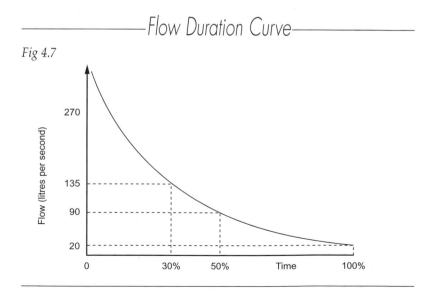

Flow Duration Curve

Fig 4.7

For the curve above, a flow of 90 l/s is exceeded for 50% of the year. This is known as the 50 percentile flow. Data in this form is particularly useful: combined with the head and other scheme characteristics, it can be used to predict the amount of energy that could be produced by a proposed scheme.

A catchment area has inbuilt storage, which slows water run-off to the stream, and keeps the water flowing when the rain has stopped. A scheme with high storage catchment will have a flatter flow duration curve, since more flow is available for more of the year. As a consequence, more energy will be generated by the scheme. A steep curve suggests a flashy catchment, where rainwater runs off quickly, the flow rises and falls quickly, and the inbuilt storage is small. This is characteristic of the following landscape features: steep slopes, exposed impermeable rock, shallow soil, and little vegetation cover. A flatter curve is due to the opposite effects. Size also effects the flow duration curve; since rainfall and local effects become averaged out over the catchment, a large catchment will tend to have a flatter curve than a small one.

The average daily flow is typically exceeded for approximately 30% of the year - equivalent to the 30 percentile flow. Commercial schemes are usually designed on the basis of a design flow equal the level of the ADF and will consequently use the 30 percentile flow. On the other hand, a typical domestic scheme will normally be designed on the basis of a lower flow than the ADF, hence the 50 percentile flow might be used as the design flow, and the design flow would be exceeded for at least 50% of the year.

Deriving flow data for your site

Flow monitoring

Analytical methods have an accuracy of, at best, ±10% (worse on very small catchments), so the ideal method of predicting the flow at the intake is to take actual flow measurements (see 'Measuring flow' below) at regular intervals (ideally daily) for a period of up to a year or more. The collected data can be compared to longer term rainfall or river flow data. For example, information can be found from the Institute of Hydrology about how the monitoring period's rainfall compared to the average expected rainfall. The long-term

average can be calculated by multiplying the recorded average
flow by a ratio based on how it compared to the expected flow.

Stream correlation

It is often the case that no stream flow data will exist for your site.
If so, it is usually possible to make a good estimate of the flow by
using the stream correlation method. Essentially, a model of the
likely flow characteristics for the 'target' stream is constructed on
the basis of analysis of the flow patterns of the nearest neighbouring stream or river for which data has been compiled.

The process is fairly straightforward, and can be conducted on
two bases:

- having no flow measurements whatsoever for the target
  stream
- having a limited number of flow measurements

The latter will give more accurate results, but the former may be
useful in the early stages of the site evaluation. In both cases, the
first job is to obtain the data of the neighbouring stream or river.

Where the correlation is to be made on the basis of having taken
no flow measurements, the data should be obtained in tabular, as
opposed to graphical, form. Then, the net annual rainfall figures
and catchment areas should be compared for both the target stream
and the neighbouring stream. After adjusting for the different
catchments and rainfall, correlation can then be carried out. The
new flow data is estimated by multiplying the known flow data by
an adjustment factor, which is created by multiplying the
catchment area and annual net rainfall level of the target stream,
then dividing the product by the product of the catchment area and
annual net rainfall level of the gauged stream.

Example

Target stream: Annual net rainfall = 2.5m   Catchment area = $3km^2$
Gauged stream: Annual net rainfall = 2.8m   Catchment area = $20km^2$
Hence, the adjustment factor   =   (2.5x3)/(2.8x20) = 0.134
Hence, your data set may look something like the following:

Data item	Flow of gauged stream	x Adjustment	Estimated flow of target stream
a	1230 l/s	0.134	165 l/s
b	1050	0.134	141
c	1080	0.134	145
d	1470	0.134	197

Computer spreadsheets can then be used to build up a quick hydrological model and assess the potential output.

This method assumes that the catchment areas have the same run-off characteristics because they are nearby. Run-off characteristics are affected by numerous factors, including soil type, slope, vegetation cover. Obviously no two catchments will have identical characteristics, but adjacent catchments can be remarkably similar. The factors affecting run-off should be considered when using this method.

Where some flow measurements have already been taken (at least one per month over a year) it is possible to make a more accurate estimate of flow. As before, the figures for the gauged stream should be obtained. Isolate the figures for the days which correspond to those on which the target stream's flow measurements were taken, then plot a graph of the two (fig. 4.8).

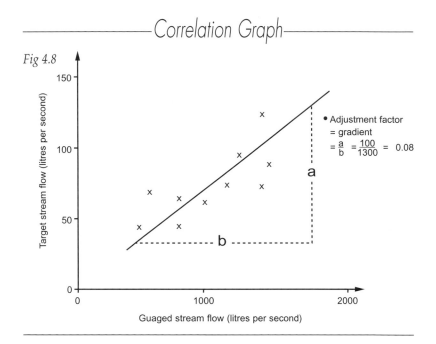

─────────── *Correlation Graph* ───────────

Fig 4.8

The gradient of the graph provides the adjustment factor, as described earlier, and a new set of flow estimates can be drawn up.

Catchment analysis

Rather than working from data collected from another stream, this method considers the factors that directly affect a catchment's run-off. In the UK, this method is used by the latest Centre for Ecology and Hydrology computer model 'Low flows 2000' which predicts the flow duration curve, given the geographical position of the catchment boundaries. It has the potential for more accuracy than the correlation method, since it does not build on other data, which may include errors, and fewer assumptions are hazarded. A good reference for this method is Low Flow Estimation in the UK Report 108, Centre for Ecology and Hydrology, Wallingford. Copies of the modelling software are prohibitively expensive, but CEH offer an analysis service. A simplified and cheaper model has been developed specifically for assessing potential hydro power sites, known as HYDrA. However, the system is not appropriate for schemes of less than 20kW in size.

On-site stream flow measurement

On-site instantaneous or 'spot' measurements of flow are essential to check the hydrological model and give a feel for the actual available flow at the site at any given moment in time. Estimation of available flow is more difficult than head measurement: because it varies constantly, flow is difficult to measure accurately. Visual estimates are not feasible since the velocity cannot be seen directly.

There is more than one approach to measuring flow, with no one method being universally superior. The best method is dependent upon budget, time constraints, the stream type, and the required accuracy. A 100kW site may justify a very accurate assessment of its catchment made by professionals using an automatic stream gauging station with data collection every 15 minutes. A domestic scheme proposed by someone with time to spare may well suit a more low-tech approach.

In keeping with this, the following methods described are commonly used for measuring flow in a natural watercourse:
- bucket method
- velocity-area method
- salt dilution method
- stage control with a gauging weir
- stage control with a natural feature

Bucket method

The simplest method is only appropriate for very small streams: the entire flow is simply directed into a container of known volume and a record is made of the time that it takes to fill.

Do the test at least three times and take the average filling time.

$$\text{Flow rate (l/s)} = \frac{\text{Bucket volume (l)}}{\text{time to fill (s)}}$$

You will need to repeat this as often as possible throughout a year. The more measurements you have, the more accurate or reliable your estimate will be.

The Bucket Method

Fig 4.9

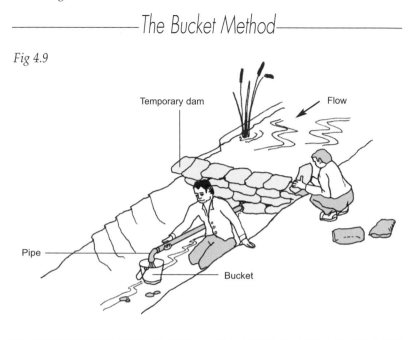

Temporary dam

Flow

Pipe

Bucket

Velocity-area method

For larger stream, a figure for the flow will be a derivation of the cross-sectional area of the watercourse and the water's velocity. The flow rate is a product of the two, hence:

Flow rate   =   velocity   x   area
(m^3/s) (m/s) (m^2)

The most common method for measuring velocity accurately is to use a current meter. This device gives a direct read-out of velocity, typically from a propeller and reed switch. Since the velocity throughout the cross-section will vary, the average velocity must be found by taking measurements at different points and calculating the average. If you use this method, choose a stretch of watercourse that is fairly even in width and depth, with a reasonable water depth, away from rapids and large rocks.

A more simple approximation of average velocity can be made by measuring surface velocity (for example, by using a float and timing the movement between two points) and multiplying this by

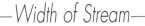

―――――――――――― *Width of Stream* ―――――――――

Fig 4.10

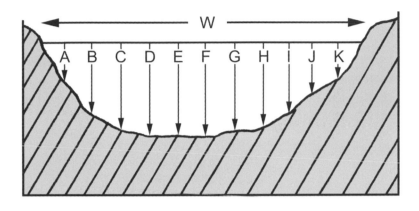

a factor of between 0.8 for a straight smooth channel, and 0.6 for a rocky stream.

The cross-sectional area can be calculated by measuring the depth at several points along a cross-section of the stream, taking the average, and multiplying by the width at that cross-sectional point.

$$\text{Cross-sectional area} = \frac{W \times (A + B + C + D + E + F + G + H + I + J + \ldots)}{x}$$

where x is the number of areas taken $(= A + B + \ldots + 1)$.

Salt dilution method

The preferred technique of the professional, the salt dilution (or salt gulp) method, provides a quick and accurate measurement of flow. The basic technique is to add a known quantity of salt at an upstream point and then monitor its progress through the water-course at a point downstream. As salt conducts electricity, its presence in the water can be gauged by measuring changes in the conductivity of the water with a purpose-built conductivity meter. On the basis of parameters that include not only the conductivity, but also the mass of salt added, the distance between the measuring points, the salt's passing time, the water temperature, and the background conductivity levels, a reasonably sophisticated meter will give an estimation of the stream flow, to an accuracy of around ±7%.

Stage control with a weir
A weir may be placed across the stream such that all the water flows through a v-notch (for low flows), or rectangular shaped (for higher flows) weir.

———— V Notch and Rectangular Weirs ————

Fig 4.11

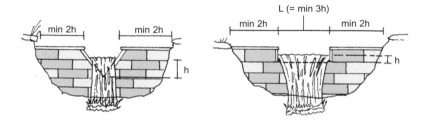

With reference to figure 4.11, the flow rate, Q, can be calculated from the height of water above the base of the notch as follows:

V-notch: $Q = 1.4h^{2.5}$ Rectangular: $Q = 1.8\ (L-0.2h)h^{1.5}$

For a small stream, a simple weir may be constructed out of wood and be totally portable. However, for maximum accuracy, a number of conditions must be met:

• All of the water should flow over the weir, not under or around it. This may be facilitated by attaching a rubber or plastic skirt to the bottom of the weir, and by carefully positioning the weir in the stream.

• The weir should be positioned vertically and at a right angles to the stream flow.

• The water should fall over the weir, and the weir should be ventilated by air underneath the crest of falling water.
• The velocity above the weir should be less than 0.2m/s, and the water should be free of eddy currents—a 'stilling pond' with a flat water surface for at least 2m upstream could help.
• The notch should be geometrically correct and should have sharp edges.

Variations
On catchments with an area of less than 3km², the ADF can vary by as much as ±50% from one year to the next; typically, however, a 10% variation might be expected. On a monthly basis, the average flow could quite easily be as high as ±500% of the ADF, and from day to day, far higher.

Floods and droughts
Flood flows are an important consideration in a hydro scheme, since this is when damage can occur to the more vulnerable scheme components, particularly the intake works and turbine house. If you do not properly consider potential flood flows, you run the risk of undersizing the intake, or locating the powerhouse within flood plains. People rarely observe the peak flood flows of a watercourse directly, since it will be raining heavily and they are usually indoors! Once a flood flow is predicted, you can calculate the level of the watercourse under these conditions. This should give you some idea of the extent of bank protection required at the intake, and of the positioning of the powerhouse.

Again, data from a nearby gauging station can be useful. However, the accuracy of flow measurements at these high flows is often poor.

A simple rule of thumb calculation for flood flows is:

Annual flood flow, Qa = 100 x ADF (or greater with a steep, flashy catchment)

While most schemes are designed with a twenty year life, the fifty year flood should be considered when siting the powerhouse and intake. By definition, a fifty year flood will occur once every fifty years, on average; of course, it may occur in the first year or not until the hundredth. The fifty year flood flow can be estimated as double the annual flood flow (200 x ADF).

Droughts are also important considerations in that power production ceases, but they are not likely to damage the scheme. Generally the turbine may be out of action for at around three to five weeks of the year because of dry conditions, although in a drought year this may increase to eight to twelve weeks. This 'down-time' can be used productively for maintenance purposes.

SUMMARY

Having completed the site assessment, you will now be in the position of knowing how the site is likely to be laid out and how much electricity it is capable of producing. The following chapters provide detailed explanation of how to get your scheme installed.

Chapter 5
Civil Works

INTRODUCTION

Once you have completed the site evaluation and decided on the basic layout of the scheme, the next stage is to consider the design and construction of the civil works. This chapter highlights the essential aspects involved in designing and developing the civil works of a typical micro-hydro scheme. For a more in-depth exploration of the design aspects of micro-hydro schemes, the reader is directed to the comprehensive *Micro-Hydro Design Manual* by IT Publications.

A great deal of thought goes into a well designed hydro scheme, and it is purpose-built for the site. Do not underestimate the importance of good design: if you put a windmill up on a poor site, you simply get less power; if you install a water turbine designed for the wrong head, flow, pipe, or intake, your problems could be far greater. There are also safety issues—particularly relating to high pressure pipelines—which are discussed in the penstock section.

The civil works required will depend entirely on the nature of the site, but are likely to include some or all of the following:
- intake
- channel
- settling tank
- penstock
- powerhouse

——————————————— *Civil Works* ———————————————

Fig. 5.1

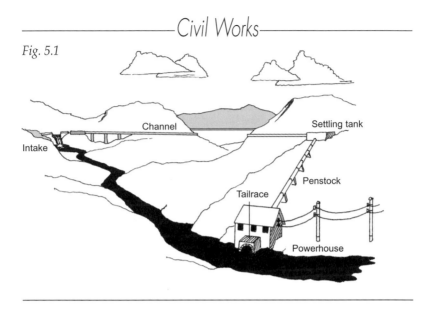

THE INTAKE

The intake is the point in the hydro scheme where a controlled flow of water is diverted from the river or stream. The intake has two main functions: to provide a consistent **quantity** of water, throughout the year, in as broad a range of streamflow conditions as possible; and to ensure water **quality** by keeping it as free as possible of sediments and debris, and by preventing the ingress of fish.

The careful design and siting of the intake is essential to the long-term reliability and success of the scheme.

Design of the intake

A **diversion weir** will often be required to seal the stream bed and to provide a stable point for the diversion of water into the scheme. Depending on the stream conditions, this may require the construction of a large, time-consuming and expensive reinforced concrete structure. Often, however, the same effect is achieved by a natural feature of the river, such as a pool, or by rearranging some rocks or boulders. The intake itself may be separate from the weir, or it may form an integral part of it.

—————————————— A Typical Intake ——————————————

Fig. 5.2

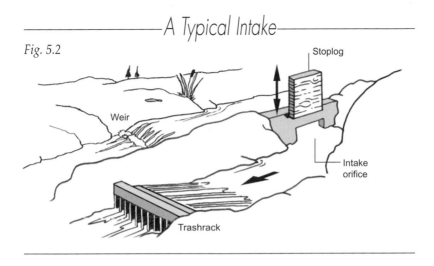

Water is diverted from the flow through the **intake orifice**. This is elevated above the river bed at a height such that the design flow is maintained at all times. If a weir is used, the intake orifice should be set back some distance such that any sediment accumulation occurs downstream. To control the flow of water, some form of **gate** will need to be incorporated: either a sliding gate, which will allow the flow rate to be adjusted, or a stoplog, used to completely stop the flow for maintenance purposes.

Take into account susceptibility to flooding, and consider the fifty year flood flow (see previous section on hydrology): the construction of **wing walls** (walls at the side of the weir to protect the banks) and/or keying into bed rock may be required. Never underestimate the erosive powers of water. Either adopt the principle of building a long lasting stone or concrete weir, or expect to replace it after major floods. (See The Micro-Hydro Power Design Manual for more details)

A small **reservoir** may be included to allow for the daily storage of water. During very dry periods when there is not enough water to produce power continuously, storage will allow the scheme to be run intermittently, at times when the power is needed. Small reservoirs are often appropriate where hydro is the sole source of power. In sizing a reservoir, consider the minimum turbine flow and a running time of perhaps six hours per day. This gives a minimum useful volume.

Reservoirs for seasonal storage are used with large schemes, but are not a part of micro-hydro systems.

Screening at the intake

To control the quality of water entering at the intake and to protect the turbine against blocking with debris, adequate screening will be required.

There is a trade-off between maintenance and cost. A few spaced steel bars or mesh are very cheap, but require regular cleaning. At the other end of the scale, maintenance-free screens are commercially available—these will keep out particles as small as coarse sand and can be left unattended. These are expensive (£500+) but usually remove the need for a de-silting chamber. Most schemes have something in between these extremes, usually a **trash rack**—which is a row of parallel tightly spaced bars.

————————*Intake Screens*————————

Intake weir with maintenance-free screen (Dulas Ltd)

Several factors need to be considered when choosing a screening method:

Cleaning: The screen should be designed for easy cleaning—inclined bars with easy access are suitable. The screen should be large enough to require, at most, daily cleaning during the worst times of year.

Spacing: The bars should be sufficiently close together to prevent the ingress of debris and fish—there should be 10mm of clear space between each bar.

Design flow: The design flow affects the area of screen required.

Area: The screen should be big enough to keep approach velocity down, to reduce blockages, and to prevent fish from sticking to it.

Velocity: As a rule of thumb, the trash rack should be designed such that the velocity of the water through it is less than 0.2m/s.

Strength: If the screen becomes completely blocked, significant forces may be exerted on it. Also, potential impacts from rocks and other solid objects during flooding need to be considered.

Freezing: Non-metallic screens can be used where there is a major risk of freezing. Otherwise, metal screens should perform quite adequately, especially if kept below the surface of the water.

A skimmer may be used in conjunction with the screen to reduce the levels of any floating debris. This may simply consist of a pole secured across the water's surface between the intake wall and the river-bank or weir.

Siting the intake

When siting the intake, consider the possible effects of flooding, erosion, and debris or sediment accumulations. These can be minimized by working in harmony with the river and giving due consideration to its natural features. Also consider the weir foundations and getting the drawn water away from the river bed.

River bed stability

The river bed should be assessed for its vulnerability to erosion; where possible, site the intake on bedrock. If this is not possible, an extended apron with wing walls or cutt off wall will be needed to prevent leakage and undermining. Concrete is favoured because it is cheap, strong, and easy to use. Note that cement contains hazardous chemicals, and care must be taken to minimize any effects on the stream. Other materials such as wood, stones, and natural features, can be used for weir construction.

Bends in the river
As a general rule, the intake should be placed on a straight section of the river. Where the intake is to be located on a bend, the outside of the bend is usually favoured—sediments tend to accumulate on the inside of bends, while the river scours the outside clean.

Rocks and boulders
Rocks and boulders can be used to construct a weir, as discussed previously, or situated such that they offer some degree of flood protection. Obviously it is important to place them where they cannot cause any damage themselves. Use natural features wherever possible to minimize any negative visual impacts of the scheme.

Accessibility
Finally, make sure the intake is accessible. The degree of accessibility required will depend on the scale of construction work required. Vehicular access can save much time, pain, and money, especially if concrete is to be used—bags of sand and cement are heavy. Additionally, the intake may need routine checks and maintenance: the more easily this can be done, the more likely it is to be done!

CHANNELS
Channels may be used as a low cost option (in terms of materials) for transporting water horizontally from the intake to the penstock. They usually take the form of a lined or unlined open canal cut into the soil. The main advantage of using a channel is the reduction in the length of the penstock, hence keeping the costs of the pipework down.

Channels are widely used in developing countries due to low labour rates. In the West, their use in micro-hydro is often restricted to very low head schemes. They are mainly limited by high labour costs coupled with the availability of low pressure plastic pipe, which usually works out cheaper to install and maintain, and is far less prone to damage from landslides and flooding.

———————————— *Channels* ————————————

Fig. 5.3

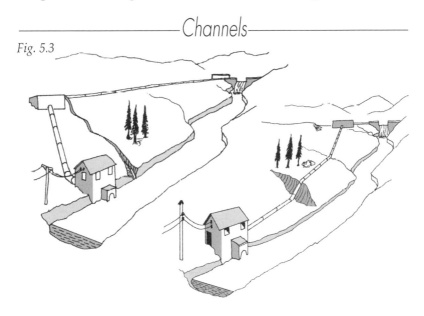

Channel with short penstock versus long penstock

There are seven main factors to be considered in the construction of the channel:

Flow: The channel must be sized for the design flow. The appropriate dimensions can be calculated as a factor of design flow, velocity, surface roughness, and gradient.

Velocity: The velocity in the channel must be limited (according to the type of lining used) to prevent erosion. Velocities in an unlined channel should, as a rule, be kept below 0.2 m/s, whereas with a lined channel there will be more flexibility.

Gradient: This affects velocity and hence erosion. For a typical unlined channel, a gradient of 1: 300 is used.

Seepage losses: If the soil is very porous, a lining is required. Where the soil is naturally sealing, for instance with clay, no lining is required. Wood, concrete, plastic, and steel may all be used for lining channels. Semi-porous channels can be used unlined, but an assessment of the seepage losses should be made.

Flooding: Consider flood levels, and the subsequent increased flow in the channel. Spillways should be installed every few hundred metres, to allow for water to be safely dumped in the event of flooding or a channel blockage. If well designed and positioned, they should prevent the channel from bursting its banks.

————————————— *Spillway*—————————————

Fig. 5.4

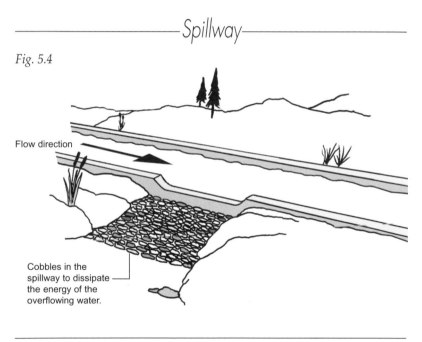

Flow direction

Cobbles in the spillway to dissipate the energy of the overflowing water.

Freezing: Slower moving water in a deep channel is less susceptible to problems, as an insulating layer of ice can form on the surface of the channel.

Cleaning: A settling tank, which can easily be flushed, may be incorporated near the beginning of the channel (see below).

SETTLING TANKS

Unless a silt-excluding screen is used at the intake, a settling tank will be required at the entry to the penstock. The purpose of the settling tank is to prevent suspended solids—mainly sand and silt—from passing through the turbine. Unless settled, these materials can act as scouring agents, causing rapid degradation to the turbine runner.

There are three main factors in settling tank design: sizing, shape and cleaning.

Sizing: The size of the tank is determined by the design flow, velocity, the size of particles to be removed, and the cleaning frequency. The basic principle is to slow the water down over a distance sufficient for the particles to fall out of suspension and sink to the bottom. The settling velocity of a particle (the vertical speed at which the particle sinks) varies according to that particle's mass. While the maximum permissible particle mass will depend on the turbine used, as a rule-of-thumb all those above 0.3mm diameter should be settled. In non-turbulent water, such particles will have a vertical settling velocity upwards of approximately 30mm per second. Hence, in a 300mm (0.3m) deep tank, the water will have to be slowed such that it takes at least 10 seconds to pass through (300/30 = 10). Assuming a volume flow of, say, 0.2m³/s, this will require a surface area as follows:

$$\text{Area} = \frac{\text{Volume flow x Settling time}}{\text{Depth}} = \frac{0.2 \times 10}{0.3} = 6\text{m}^2$$

Shape: Generally the tank should taper out to a width of approximately 5–15 times that of the channel. Sharp edges should be avoided, as these can lead to localized high velocities and turbulence, which stir up the silt. The depth of water in the tank should be maintained at a level at least equal to that of the channel that has served it. Taking account of the settling of solids, the total depth to the bottom of the tank will therefore consist of two components:

• **Channel depth:** this is the minimum depth of the water flowing through the tank when it is full of solids.

• **Collection depth:** this is the depth to which the tank floor slopes below the channel depth. It is the section to be filled with solids.

————————Settling Tank Dimensions————————

Fig. 5.5

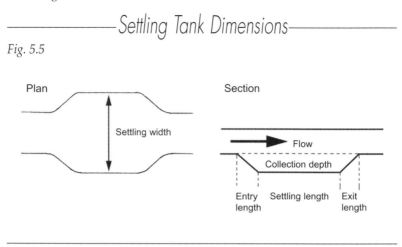

Cleaning: The tank should be designed such that it can be easily flushed from a large outlet—probably a plug built into the base of the tank. When the plug is opened, the flushing action sluices the solids away with the water.

The frequency with which the tank will need to be emptied depends on its size and design in relation to the turbidity of the

————————Settling Tank Design————————

Fig. 5.6

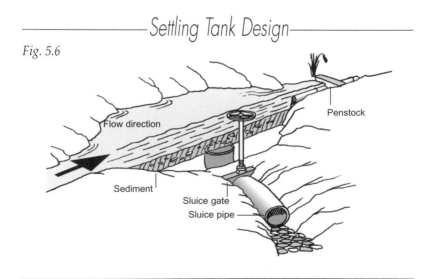

water: the greater the collection depth and volume, the less frequent the cleaning required. Turbidity is a measure of the level of suspended solids in the water (measured in kg/m³), and will vary not only from location to location, but more particularly from season to season. A site where the tank normally requires cleaning on a monthly basis may require cleaning on a daily basis during flood flows, owing to the increase in turbidity.

PENSTOCKS
The penstock is the pipe that conveys the water under pressure to the powerhouse and then the turbine. It must be able to withstand not only the high pressures associated with head, but also the even higher pressures that would be caused in the event of a blockage. The penstock is often one of the most expensive elements of a hydro scheme, so particular attention should be paid to its design.

The Penstock

Fig. 5.7

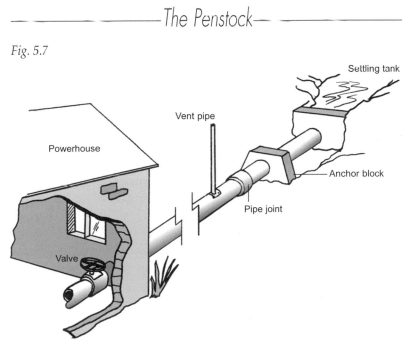

Components of a penstock assembly

Main design considerations

The size of the pipe (its wall thickness, and internal and external diameter), the material from which it is fabricated, and how the pipe is to be laid are all factors that will be determined by the following imperatives:
- to minimize cost
- to maximize the available head, and minimize any head losses
- to withstand the anticipated maximum water pressure
- to last the lifetime of the scheme with minimum maintenance

The following factors will need consideration:

Static head (pressure)

Static head determines the wall thickness of the pipe to be used. The higher the head, the stronger the pipe must be; strength, for any given pipe material, is achieved by increasing the wall thickness. The pressure in the pipe increases with head: for every 10m increase in head, the pressure in the pipe rises by one bar (100 kPa).

Surge pressure

In practice, the water pressure can rise above the static pressure head—sometimes by 50%, or more—due to surge pressures. These occur when the flow of the water through the pipe is suddenly halted—by a blockage, for instance, or if a valve is closed too quickly. The effect is very similar to the 'water hammer' effect in plumbing. Unless sufficient precautions are taken, the consequences can be disastrous.

Allowance for this must be made by using overrated pipe. If possible, the pipeline route should take the possibility of catastrophic failure into account.

Design flow rate

Design flow rate determines the internal diameter of the pipe to be used. The higher the design flow, the larger the diameter of the pipe required. Too small a pipe will result in unacceptable head losses.

Head loss

Energy losses occur during transmission of the water through the penstock, causing a reduction in the effective head. The difference between the real head and the effective head is the 'head loss'. For a given pipe material and design flow rate the head loss is a function of pipe diameter—one of the main reasons for schemes to fail is the installation of a pipe of too small diameter. However, a certain degree of head loss can be tolerated, typically around 5–15% of the gross head. As the degree of acceptable head loss determines the diameter of the penstock pipe, a higher acceptable head loss allows for a smaller diameter pipe to be used.

There are two causes of head loss. One is the roughness of the internal surface of the pipeline (often measured in terms of the 'k-factor'); the other is turbulence arising from bends, sudden contractions, and valves in the pipeline (although this is less significant than roughness). Head loss is proportional to the square of the flow; hence, doubling the flow quadruples the head loss. The actual calculation is fairly complex, and manufacturers of piping provide tables or graphs of head loss versus flow, such as figure 5.8.

——————*Typical Head Loss Versus Flow Graph*——————

Fig. 5.8

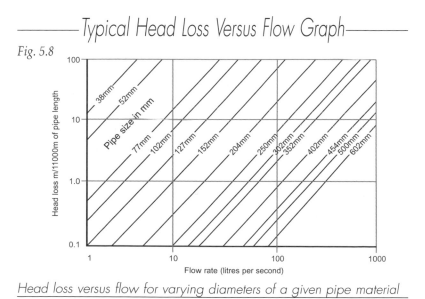

Head loss versus flow for varying diameters of a given pipe material

Remember such graphs apply to new pipes. The internal surface of any pipe will degrade over time and the k-factor (and therefore the head loss) can increase dramatically.

In terms of roughness, plastic or epoxy-painted pipes are smoothest, whilst concrete and old cast iron pies are the roughest. This means that it is possible to use a smaller diameter plastic or epoxy painted pipe than concrete pipe for a given acceptable head loss.

The head loss is more critical on low head schemes, as the head loss is proportional to flow and pipeline length, and not related to the scheme's overall head.

Pipe degradation

There are two areas of degradation: deposit build up and corrosion. Where the turbine flow is provided by a catchment with peat bog, for instance, a peaty deposit builds up on the inside of the pipe, increasing head loss. The pipe should be designed for this, or a facility for easy cleaning made. Mild steel pipes are particularly susceptible to corrosion, and allowance must be made for this in the specification.

Pipeline route ground conditions

If the ground is very rocky, it may not be possible to bury the pipe, and the pipe may have to be installed above ground. In such circumstances, the pipe will be more vulnerable to damage and a robust material, such as steel or ductile iron, may have to be used.

Material selection

The following are some of the materials that can be used for penstock pipes in micro-hydro schemes:
  • unplasticized polyvinyl chloride (PVC-U)
  • polyethylene
  • glass reinforced plastic (GRP)
  • reinforced concrete
  • mild steel
  • cast and ductile iron

Each has differing strength, cost, weight, and surface roughness. All of these factors will need consideration.

Plastics

For heads under 90m, PVC-U is often favoured for its lightness, ease of juncture, low friction factor, longevity (because it is resistant to corrosion), and low cost. However, PVC-U pipe must generally be buried, since it is vulnerable to impact damage and degrades in sunlight. Further, if part of your rationale for installing a micro-hydro system is an ideological one based on lessening your environmental impact, PVC-U is often considered to be the least environmentally-friendly option and alternatives should be sought where possible.

Polyethylene has all of the advantages of PVC-U as well as being sunlight resistant, more flexible, and impact proof. The piping is similar in price up to 150mm—after which point it becomes slightly more expensive—and is the prefered choice of the UK's water industry. Polyethylene is generally considered to be the environmentally-friendly alternative to PVC.

GRP (fibre glass) can be the cheapest plastic option for large diameters (more than 500mm) and high heads (up to 160m). It has many of the advantages of the other plastics. However, GRP is heavier and more fragile: installation must be done carefully, particularly on medium and high head sites.

On particularly high head sites, the pipeline can be divided into two sections, beginning with plastics in the first section, then changing to a stronger material, such as steel, to cope with the higher pressures further down.

Mild steel, cast, and ductile iron

Steel and iron pipes tend to be heavier and more expensive at lower pressures, but equal in price (whilst still heavy) for heads over 100m. The chief advantage of these pipes is that they are very robust, hence generally the best option for above-ground use. Steel is very flexible for installation, and bends can be made up on site and welded in-situ.

Steel and cast iron (which is often found on abandoned schemes) is generally rougher inside compared to plastic, so a larger diameter is required to compensate for this extra resistance. Unless concrete lined or regularly painted, steel and cast iron will become pitted inside; this increased roughness causes additional head loss.

Reinforced concrete

Since concrete has nominal tensile strength, it is not an obvious choice for pipelines. Although it is not often used these days, it was a favourite option in 6" and 8" sizes, up to 50m head, between the 1930s and 60s. It has similar benefits to steel and iron, without the corrosion problem.

Laying the pipework

Jointing

Numerous methods can be used for joining the sections of pipe together. Depending on the material, this will range from gluing and welding to push-fit connections. The pipe manufacturer will usually provide information on this.

Expansion

Pipelines can contract and expand depending on the pipe material. Huge forces can be exerted on the joints, particularly by exposed steel pipes. In such cases expansion joints must be included to allow for this movement. Pipelines joined using mechanical sockets have built-in expansion joints at each connection, and the pipe should be pushed to 5mm from home in the sockets.

Anchor blocks

Since the surface area on the outside of a pipe bend is larger than on the inside, the internal pressure generates a force that tends to move the pipeline. The same applies to an end cap. The forces involved can be very high, and in order to restrain the pipe it is often essential to place an appropriately sized anchor block at each bend and at the pipe end at the turbine house. This is simply a block of mass concrete built around the pipe such that the pipe is 'anchored' to the ground. This is obviously safety critical. See MHDM for details on design.

Burying

This is often a requirement, for protecting fragile pipes (such as those made of pvc) and aesthetic purposes (a consideration in planning permission). When burying, care must be taken that the pipe is well supported, and that no sharp objects will damage the pipe once it is covered. Once the pipeline is placed in the trench, test it for leaks, if possible, (see chapter 9) before covering it over. Certain pipes require pea gravel blinding to ensure protection. Use markers to identify the route as a buried hazard.

————Section of Penstock Prior to Burying————

Richard Langley

Costs

Table 5.1 gives indicative prices (U.K., 2004) for new pvc-u pipe, with integral push fit connections, in sizes commonly used in domestic systems (on the basis of a purchase of 100m+). As a rule of thumb, a bend costs the price of 4m of pipe, whilst a valve costs around the price of 10m of pressure pipe.

TABLE 5.1

Pressure rating	Diameter (mm)	Cost(£/m)
Up to 20m head	100	3.00
Up to 60m head	100	7.20
(class C pressure pipe)	150	14.40
	200	24.00
	250	36.00

Valves

More often than not, a valve will be required at the end of the penstock, before entry to the turbine. This allows the water supply to be turned on and off easily without draining the penstock. Common valves include Gate, Butterfly, Spear and Globe.

The Gate valve

The gate valve contains a circular gate that moves down vertically across the pipe in order to stop the flow. These are recommended for the main inlet valve as they have the advantage of being hard to close quickly.

The Butterfly valve

This valve contains a circular gate that is shaft-mounted within the water flow. Rotating the gate by a quarter of a turn opens the valve; a further quarter turn closes it. They can be used on high-head sites without any problem, but caution should be taken to ensure that the valve is not closed off too quickly as water-hammer would ensue.

The Spear valve

Often used with the pelton and turgo turbines, and essential when used in combination with a mechanical governor (see chapter 6), the spear valve allows for continuous flow regulation (it is not simply 'open' or 'closed'). It consists of an orifice in and out of which a spear-shaped horizontally shaft mounted head moves. The gap through which the water flows changes constantly to regulate the water flow.

The Globe valve

The globe valve contains a hollowed spherical gate opened by a quarter turn rotation – caution should be taken to avoid water-hammer as with the butterfly valve.

THE POWERHOUSE

The final link in the civil works chain is the powerhouse, the building at the end of the penstock that houses the turbine and electromechanical equipment. Its main function is to protect the

—————————— *Valve Types* ——————————

Fig. 5.9

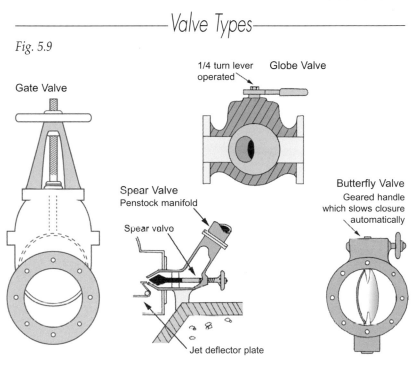

Gate Valve

1/4 turn lever operated

Globe Valve

Spear Valve
Penstock manifold

Spear valve

Jet deflector plate

Butterfly Valve
Geared handle
which slows closure
automatically

Gate, Globe, Butterfly and Spear valves

components from the elements while allowing easy access for maintenance purposes. There are several factors to consider:

Design: Since the turbine house may require planning permission, the design may have to meet local planning requirements: it should blend in with the surroundings and be environmentally sensitive.

Foundations: A strong reinforced concrete foundation is usually required to support the machinery.

Flooding: The fifty year flood flow should be considered and the building located such that it is not in a flood plain.

Ventilation: Heat is produced by the alternator; it needs to be kept cool.

Lighting: Often overlooked, natural lighting is worth considering for times when the turbine is undergoing maintenance, as well as an electric light for adjusting inlet valves at night.

———————————*A Typical Powerhouse*———————————

Dulas Ltd.

Overhead load bearing beam/hooks: In conjunction with lifting equipment, this can help during maintenance.

Noise: If noise may be a problem, avoid windows and insulate the roof, but bear in mind the need for adequate ventilation.

Size: Allow for easy access to the turbine valve(s), and consider turbine size when sizing entrance. Typical sizes range from just 1x1x1m for a turbine of less than 1kW, to 3x3x2m for 3–25kW, to perhaps 5x6x3m for a 100kW site.

Costs

The costs will obviously depend wholly on size and design. The size will vary enormously, depending on the size of the turbine to be housed, the materials used to construct the powerhouse, the extent of the foundations required, etc.

The Civil works usually account for around 70% of scheme costs.

Chapter 6
Electro-mechanical Components

INTRODUCTION

The final part of the previous chapter described the powerhouse. This chapter describes what to put in the powerhouse in order to convert the power of the water into electricity, and how to transmit the electricity from the powerhouse to where it is needed.

TURBINES

The turbine converts the potential energy of the water into rotating shaft power. The selection of the best turbine for any particular hydro site depends both on the site characteristics—mainly head and flow —and on the generator speed required.

As described in chapter 2, turbines can be divided into two main groupings, depending on the way they extract energy: **impulse** and **reaction**.

The major feature common to all turbines is the **runner**, the rotating element (usually wheel shaped) that converts the potential energy of the water into shaft power. An impulse turbine's runner operates in air and is spun by a jet of water emerging from a nozzle. The runner is designed such that a maximum of the jet's momentum is converted to rotational shaft power. After contact with the runner's blades, the water should have lost its energy and therefore drop down into the tail-race.

Impulse turbines are generally used in high head, low flow applications. As a rule, impulse turbines are fairly inexpensive owing to their relative simplicity. This, combined with good relia-bility in a range of head and flow conditions, and ease of access for maintenance, means that they are often the cheapest option for small schemes. Also, the pipeline and valves are more manageable because of the lower flow rates required.

A reaction turbine's runner is submerged in the flow, and is housed in a pressure casing. The runner and casing are carefully machined so that the clearance between them is minimized, and all of the water passes through the runner. The runner blades are profiled so that pressure differences across them impose 'lift' forces—like those imposed on aircraft wings—which causes the runner to turn.

Reaction turbines have a draught tube, which is a tapered tube that extends from the casing around the runner, down to beneath the lower water level. In effect, the draught tube increases the net head: during operation it fills with water, providing a suction head which adds to the head above the runner. Reaction turbines are used in high flow, low head micro-hydro applications.

Water-wheels and pumps-as-turbines are not included here, but are discussed in chapter 10.

Common impulse turbines

Pelton wheels
These are the best option for high head sites as they are efficient and have a long life. They are ideally suited to the high heads and low flows obtained from steep mountain streams. The runner is a wheel with a set of cups mounted around its rim; it is driven by a jet of water hitting each cup in turn.

Pelton wheels can be adapted to work with higher flows and therefore lower heads. Some extra flow can be accommodated by adding more than one jet to the runner (up to 6) to make a multi-jet Pelton. This has brought the minimum head down to around 20m.

Turgo turbines
Turgos are suited to heads of 10m upwards. They are renowned for their efficiency, reliability, and long life. However, they require expensive casting so they are not widely used, although they were popular in the past, and many are to be found on abandoned sites.

——————————————*Pelton Wheel*——————————————

Dulas Ltd

In shape, the runner is similar to half a pelton wheel, with the water entering from one side and exiting from the other. The Turgo has the advantage over a Pelton wheel of being able to take more flow, but the disadvantage that a side thrust is exerted on the bearings.

——————————————*Turgo Turbine*——————————————

Fig 6.1

Crossflow turbines

Also known as the 'Mitchell' or 'Banki' turbines, these have recently come into widespread use. One reason for their popularity is that they are relatively easy to make (see the section on 'Building your own Turbine'); many developing countries have workshops that build crossflows for local use. Crossflow turbines are also popular because they can be run over a wide range of heads—from 3–60m. In addition, while they exhibit a relatively low maximum efficiency, they do maintain a reasonably high part-flow efficiency.

──────────────── *Crossflow Turbine* ────────────────

Fig 6.2

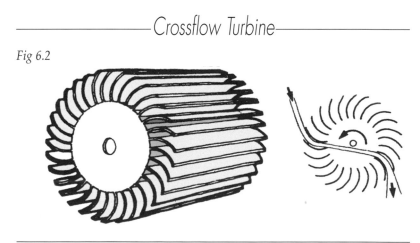

Common reaction turbines

Propeller turbines

A propeller turbine closely resembles a ship's propeller, except that the force of the moving water causes the blades to rotate, rather than vice-versa. They can be installed in several different ways, often being built into the weir or dam used to develop the head. Because of the amount of water being dealt with, the machine tends to be large. This, combined with the need for accurate machining of the blades, means that propeller turbines tend to be expensive. However, they do have the advantage of being available for heads as low as 1.5m. An even more expensive option is the 'Kaplan', which is a propeller where the angle of pitch of the blades

Propeller Turbine

Fig 6.3

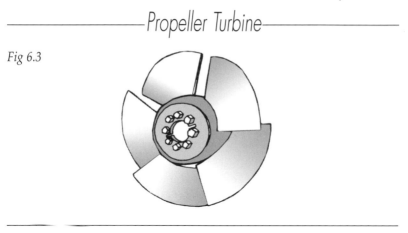

is adjustable. This allows the turbine to run efficiently at lower flow rates.

Francis turbines

While these are the most commonly used machines in large installations, they are difficult to cast and hence expensive. This has meant that there have been few applications for small versions, though recent 'fabricated' models have been available that are built

Francis Turbine

Fig 6.4

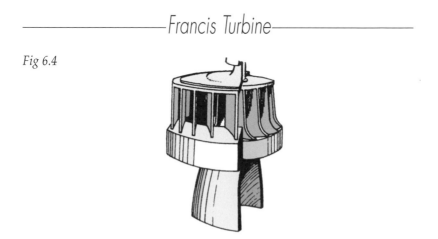

much more cheaply from forged and welded steel. Minimum operating heads are around 2.5m. However, they are prone to pitting from small stones, which can greatly reduce efficiency.

Turbine selection
Water power can be harnessed in a variety of different conditions, from very low heads to high heads. To accommodate these conditions, the various types of turbines described above were developed. As a guide, reaction turbines are suited to high flow, low head applications, whilst impulse turbines are suited to higher head, low flow applications. There is some overlap between the different turbine types, particularly with the crossflow turbine which can operate over a large range of head and flows. Figure 6.5 serves as a rough guide to turbine selection, given known head and flow.

————————Turbine Type Vs Head and Flow————————

Fig 6.5

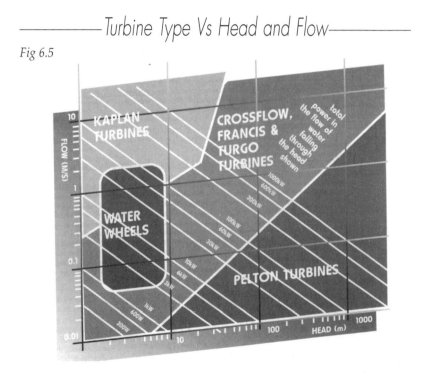

In selecting the turbine, the main factors to consider are efficiency (over a range of flows), cost, and maintenance. All of the turbines have peak efficiencies of 70–90%; impulse turbines tend to have a better efficiency than reaction turbines over a wider range of flows.

Speed matching

For maximum efficiency, the turbine must operate at the correct rotational speed for the pressure head under which it is operating. The relationship between speed and head depends on the design of the turbine. For impulse turbines, the rotational speed of the runner should be just under half of the speed of the jet of water striking the buckets. For a reaction turbine the equation is less simple, and depends on the blade angles of the particular turbine.

To simplify the process of speed matching, a table of specific speed ranges has been devised (see below). This relates a numerically defined, and for our purposes unitless, range of conditions, under which each type of turbine will operate effectively, to the following parameters:

- Intended design power, P (kW)
- Head, H (m)
- Required rate of rotation of the turbine shaft (shaft speed), n (rpm)

It functions in accordance with the following equation:

$$\text{Specific speed, } N_s \quad = \quad \frac{n \times (P \times 1.4)^{0.5}}{H^{1.25}}$$

The required rate of rotation, n, will depend on the type of generator used and whether or not it is driven directly by the turbine (generator types and drive systems are covered later in the following two sections). In order to produce 240V ac at 50Hz, most generators will need to be spun at 1500rpm. With a direct drive system this will require a shaft speed of 1500rpm; where gears or belts are used between the turbine and generator, the shaft speed may be much lower.

Example

At CAT, one of the water turbines operates under the following conditions:

Head: 30m
Power: 3.5kW
Generator speed required: 1500rpm

Assuming to start with that a direct drive is to be used (thereby making n = 1500rpm), the specific speed can be calculated as follows:

$$\text{Specific speed,} \quad Ns = \frac{1500 \times (3.5 \times 1.4)^{0.5}}{30^{1.25}}$$

$$= \quad 47$$

Table 6.1: Specific Speed Ranges

Pelton	12—30
Turgo	20—70
Crossflow	20—80
Francis	80—400
Kaplan	340—1000

From the specific speed ranges table, this would suggest that the Turgo or crossflow might be the most appropriate in these circumstances.

However, in reality a Pelton wheel has been used with a 2:1 ratio belt drive instead of a direct drive system. The use of the belt has reduced the required turbine shaft speed, n, to 750rpm, thus halving the specific speed to 23.5—well within the Pelton wheel's range.

Costs

Turbine costs vary massively, depending on the type of turbine, the manufacturer, and whether or not the unit is new or reconditioned. Cost does not necessarily correspond to quality. Since turbines are generally hand-made, costs are inevitably fairly high.

Some rough guideline prices (2004) for new combined turbine and alternator sets are given in Table 6.2.

Table 6.2

Turbine type	Power range	Cost
Pelton	Up to 1kW	£625—3,125
	1—10kW	£1,875—15,00
	10—50kW	£6,250—62,500
	50—100kW	£25,000—150,000
Francis	Up to 30kW	n/a
	30—50kW	£37,500—100,000
	50—100kW	£62,500—250,000
Crossflow	Up to 1kW	£625—4,375
	1—10kW	£2,500—25,000
	10—50kW	£12,500—87,500
	50—100kW	£37,500—187,500
Propeller	Up to 1kW	n/a
(fixed blade)	1—10kW	£3,750 —25,000
	10—50kW	£10,000—62,500
	50—100kW	£37,500—200,000

The cheapest turbines are available from agricultural style businesses, or developing countries such as China.

Building your own turbine

Unlike most other types of turbine, the crossflow can be fabricated using basic workshop machines, although its service life and reliability may be below that of the conventional options. Even if you don't have access to a lathe or mill, most of its construction can be carried out with basic tools, cutting gear, a bending press, and an arc-welder. The machining will need to be done in a larger workshop. The overall initial cost saving will be greatest where the machine is intended for use at a low head site, but the overall life will be limited by wearing of the blades.

It is also be possible to build most other types of turbine, at least in part. The casing and jets can be fabricated; the runner may have to be purchased separately as it needs to be precision cast. Of course, you may wish to cast your own runner. In the case of the Pelton for instance, this need not be too complicated (indeed, here at CAT, an annual course is run for those wishing to do just that!).

Remember that careful design is always essential to ensure correct operating characteristics and efficient operation.

Designing and building any turbine is a sizeable undertaking. Assuming you are a skilled turner and fabricator with all the required equipment, and mathematical skills to carry out the design process, the work may require up to four weeks' labour.

For low heads, water-wheels (see chapter 10) can also be a good option for the keen would-be turbine builder—despite their drawback of low speed, water-wheels can cope with a wide range of flows. While careful design is essential, less precision machining and fabrication is required.

DRIVE SYSTEMS

The drive system transfers power from the turbine shaft to the generator shaft. It may also have the function of gearing. This allows the rotational speed of the turbine to be synchronized with that required by the alternator (except for rare cases where the two are the same and a direct drive can be used). Because small turbines tend to use standardized runners, gearing is often required; only with larger, 50 kW+ schemes are the runners purpose-built to facilitate direct drive.

Power transmission, lifetime, cost, and maintenance are the main factors to be considered when designing the drive system. Possible options for the drive system include:

- direct drive
- belt drive
- gearbox

Direct drive

Directly coupled drives are the best option where possible, since they are nearly 100% efficient in transferring power from the turbine shaft to the generator. They are maintenance-free, do not wear appreciably, and do not exert any lateral stresses on the bearings of the generator shaft.

However, their use is limited to applications where the turbine and generator can run at the same speed. Often this is not possible, due to the need to use standardized runners and alternators; although higher power installations have more flexibility in the matching of these components. It also makes it impossible to adjust speed matching if you got your sums wrong.

Belt drive
Belt drives offer a highly efficient means of catering for the differing optimum speeds of the turbine and the generator. To work efficiently, belt drives require some degree of maintenance— mainly to ensure that they are kept clean and dry, properly tensioned, and correctly aligned. Three different belt types are worth considering:

• Flat belts: Very efficient (95–98%) and quiet, but must be kept highly tensioned—leading to large bearing loads. Good alignment important. Gear ratios of up to 5:1 are possible.

• 'V' belts: Robust drive with wide tolerance of alignment and tension adjustments, but limited efficiency (90–95%), particularly if set up incorrectly. Gear ratios of up to 5:1 are possible.

• Toothed belts: Very efficient (95–98%) and only a low belt tension required. Possibly the best option for small drives (less than 15kW), but less widely available than flat and 'V' belts, more expensive, and with a slightly shorter service life. Good alignment is important. Gear ratios of up to 7:1 are possible.

——————————————— ——*Belt Types*————————————————

Fig 6.6

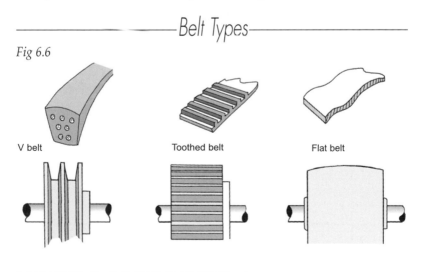

V belt   Toothed belt   Flat belt

Gearboxes

Gearboxes can allow for gear ratios of 10:1 through to 50:1. They are generally more noisy, expensive, and less efficient than belt drives, and are not recommended for most small systems. An exception to this rule may be in the case of water-wheels, where a gearbox may be the only viable option for the high ratios required for electricity generation.

ELECTRICAL EQUIPMENT

Because of the reliability of the resource, most micro-hydro schemes are configured to generate 240V ac electricity directly; only the very smallest systems (<0.5kW) will be used for dc generation (usually for battery-charging purposes). This inherent simplicity puts micro-hydro at an advantage over other small-scale renewables—normal electrical appliances can be used directly from the power generated.

ac electricity

A basic understanding of some aspects of ac electricity will be required, if only to make sense of the terminology associated with this aspect of a hydro scheme. A fuller understanding can be gained by reference to more specialist texts, such as the latest edition of Hughes' *Electrical Technology* published by Longmann.

Alternating current electricity is generated by a machine, usually an alternator, in which a coil rotates in a magnetic field; the voltage and current produced vary sinusoidally with time as shown in figure 6.7.

Voltage/Time Graph

Fig 6.7

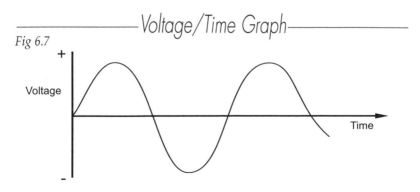

Current/Time Graph

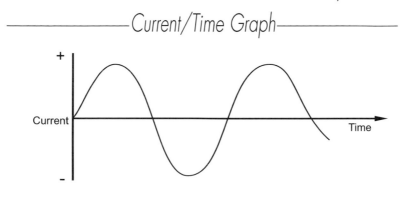

The number of cycles per second is known as the **frequency** and is measured in Hertz (Hz); in the UK and Europe, most electrical appliances are designed to operate at 50Hz.

When the voltage and current cycle together, along the same sine wave, they are said to be **in-phase**. When the voltage and current waveforms in an ac power system are in-phase, the total power available can be described by the power definition:

Power [P] = Current [I] × Voltage [V]

However, in some circumstances there is a lag between the two and the current trails the voltage, or vice versa. In this case, they are **out-of-phase**.

Voltage and Current out of Phase

Fig 6.8

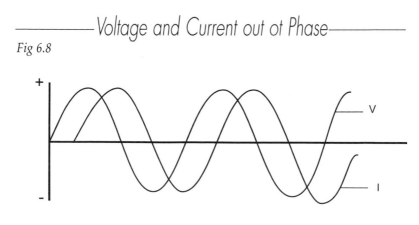

Because of the phase difference, there is a reduction in the useful power available, and the power definition becomes:

Power [P] = Current [I] x Voltage [V] x **Power factor [Pf]**

The power factor takes account of the lag (or lead) between the current and voltage waveforms in an ac power system. The phasal difference between current and voltage in an ac system is dictated by the power factors of that system's various load components:

- Purely resistive loads (where the current and voltage move in harmony), e.g. incandescent bulbs, bar and immersion
  heaters: Pf = 1.0
- Inductive loads (where the voltage waveform leads the current), e.g.   fridge, power tool, and washing machine
  motors: Pf = 0.4 - 0.7
  fluorescent lights: Pf = 0.5 - 0.7
- Capacitative loads (where the current leads the voltage). If the loads are mainly inductive, it may be possible to equalize the overall power factor by introducing capacitors across the circuits into which the inductive loads are connected.

The capacity of a new generator will be rated either in **kW** (power rating) at a given power factor, or in **kVA** (volt-amp rating). Conversion between the two is as follows:

Volt-amp rating (kVA)   = Power rating (kW)
                                                       Power factor

Hence, a generator rated at 4 kW at a 0.8 power factor has a volt-amp rating of 5 kVA.

The kVA rating is also known as the **apparent** power, and the kW rating as the **real** power. It is important to make this distinction; in practical terms a difference between the two means that a higher current is required to provide a given power than might otherwise be expected.

Generators

There is a choice between two basic sorts of generator. Both produce ac electricity, but one generates at a constant frequency (synchronous), and the other at a variable frequency (asynchronous).

A **synchronous** generator (or alternator) is similar to the type of machine conventionally used in large power stations, only scaled down for micro-hydro use. Compared to asynchronous machines, they have the advantage of easily producing a mains type of electricity; the frequency generated is directly linked to the turbine speed. However, they can be quite expensive and may need to be protected from running over-speed.

An **asynchronous** generator (or induction generator) is often simply an electric motor driven in reverse to produce electricity. They are cheaper, more robust, and more widely available than synchronous machines. However, their frequency is dependant not only on shaft speed, but also on load. Because their frequency is prone to fluctuation, their application is limited—problems could occur when using sensitive electrical goods such as hi-fi equipment, TV and video, computers, etc. The situation can be improved with an induction generator controller (IGC) which improves frequency control; however, this increases costs.

It is generally recommended that, unless money is very tight or the scheme is particularly small, a synchronous alternator is used. In most cases the cost savings to be had from choosing an asynchronous generator will be negligible when compared to the overall scheme cost. Let us examine some of the key considerations in alternator selection.

Automatic Voltage Regulator (AVR)

As the load on a synchronous generator increases, there can be a tendency for the output voltage to fall. An AVR is usually necessary to maintain the voltage at a constant level; most modern generators now incorporate an AVR as standard. AVRs work best in diesel-driven generators; with hydro they tend to be less reliable, owing to variations in temperature and moisture levels, fluctuations in shaft speed, and vibration. The life expectancy of the AVR may be improved by removing it from the generator and enclosing it in a box on the wall of the powerhouse.

Bearing loadings
The most readily available small alternators are those used with
diesel generators and mounted in line with the engine. With a belt-
driven hydro scheme, the alternator is mounted to the side of the
turbine. If the alternator is not designed to operate in such a
position, there is a risk of excessive radial loading to its bearings.
(If the alternator is directly coupled to the turbine shaft, there
should be no problem.) Therefore, it may be necessary to use an
alternator with up-rated bearings, or install extra bearings as
shown in figure 6.9.

——————*Belt Drive Turbine with Extra Bearings*——————
Fig 6.9

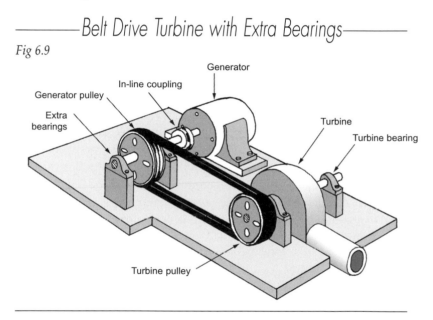

Number of poles
Most alternators are '4-pole'. The number of poles determines the
shaft speed required to generate electricity at a given frequency.
For a frequency of 50Hz to be generated by a 4-pole alternator, a
shaft speed of 1500rpm is necessary. The shaft speed is related to
number of poles as follows:

Shaft speed   = $\dfrac{120 \times \text{frequency}}{\text{number of poles}}$

The greater the number of poles, the lower the shaft speed for the same result. Some alternators are built with 6 or even 8 poles, and these require shaft speeds of 1000 and 750rpm respectively. The positive consequence of this is that less or no gearing is needed between the alternator and the turbine shaft. Unfortunately, cost usually limits such alternators to large-scale schemes.

Single or three-phase
For systems of up to 5–10kW, it is usual to use a 'single-phase' alternator; above this size range, 'three-phase' units become more common and less expensive. A three-phase unit produces ac electricity as three separate sine waves, or phases, each a third of a cycle apart from the other. A three-phase alternator may be used to supply a single-phase load.

Oversizing
It usually makes sense to purchase an alternator with at least 25–50% higher kVA rating than the maximum expected load. This oversizing takes account of power factor and the difference between real and apparent load. A further advantage is that efficiency should improve (alternators tend to be more efficient the larger their capacity). If using an electronic load controller (see below), the alternator should be oversized by 60%.

Insulation
The alternator should be specified with 'tropicalized windings' to protect against corrosion in damp conditions.

Governing
The governor controls the speed at which the alternator spins, thus ensuring that the frequency and voltage are constant—essential for the proper functioning and longevity of most electrical appliances—and that the turbine runner spins at its design speed, thus ensuring maximum efficiency.

Flow control
Until recently, flow control was the usual method of governing. With flow control, the flow of water to the turbine is mechanically

regulated such that the input water power matches the demanded power output. Such systems have moving parts which require maintenance, and they are slow to respond to variations in load.

Electronic load control
In 1979, the electronic load controller (ELC) was developed. This was a major advance on the existing mechanical flow controllers for stand-alone systems. The ELC keeps the alternator's speed constant by providing a constant electrical load on a generator in spite of changing user loads. This permits the use of a turbine without flow-regulating devices and their associated governor control system. The ELC maintains a constant generator output by supplying a secondary 'ballast' load with the power not required by the main load. The ballast load is ideally suitable for uses that can be run at power levels below their full rating, such as space and hot water heaters. Having no moving parts, ELCs are usually maintenance-free and very reliable.

When a house is already connected to the mains, the ELC can be used in conjunction with an 'auto-changeover', which allows the household's principal load to be run off the hydro generator until its capacity is exceeded, at which time the loads are changed over to the mains. The total output of the hydro is then directed to the ballast.

TRANSMISSION
Once the power has been generated, it must be transmitted as efficiently as possible to wherever it is to be used. The cable length should be kept to a minimum in order to minimize power losses and costs—cable costs can represent a sizeable portion of the overall cost of the scheme. The main factors to be considered include the following:

• Power to be transmitted: The higher the power output, the greater the cross-sectional area of cable required (cable conductor size is related to the current flowing).

• Voltage: Voltage affects the grade of insulation required and the current required for a given power (the higher the voltage the lower the current), and hence cable size.

• Transformers: If you are transmitting long distances (more than 700m), it becomes economical to increase the voltage through the use of transformers. This reduces the current flowing through the cable and therefore keeps cable size to a minimum.

• Underground or overhead: Underground cables are generally more expensive than overhead cables at powers in excess of 20kW. However, they are less visually intrusive, more reliable, maintenance-free, and can be cheaper to install in certain conditions. Underground cables should be buried to depths as specified by the appropriate regulations (typically 450mm deep, with warning tape at 150mm).

• Volt drop: Since a cable offers resistance, there is an associated voltage drop and power loss with current flow. The power loss is due to simple resistance and reactance, which relates to the cable's capacitance. For bunched conductors, such as underground armoured cable, the former need only be considered. Generally, cable is designed for a maximum 5% power loss, i.e. a volt drop. The volt drop can be calculated as follows:

$$\text{Volt drop (v)} = \frac{0.004 \times \text{cable length (m)} \times \text{current (A)}}{\text{cross-sectional area of the cable (mm}^2)}$$

Having established the acceptable volt drop, it is possible to calculate the minimum cross-sectional area of the cable.

Example

Consider a single-phase 5kW, 240V ac system, with a 150m underground steel wire armoured copper transmission line. Assume a power factor of 0.7. The minimum cable size required to ensure a volt drop of no more than 5% can be calculated as follows:
Firstly calculate the current:

$$I = \frac{\text{Power}}{V \times \text{power factor}} = \frac{5000}{240 \times 0.7}$$
$$= 30 \text{ A}$$

Now, rearranging the volt drop equation,

$$\text{area of the cable (mm}^2) = \frac{0.004 \times \text{cable length (m)} \times \text{current (A)}}{\text{Volt drop (v)}}$$

and allowing for a volt drop of 5%: $240 \times 0.05 = 12V$,

$$\text{area of the cable (mm}^2) = \frac{0.004 \times 150 \times 30}{12} = 15 \text{ mm}^2$$

Hence the minimum cross-sectional area of the cable equals 15mm^2. The nearest standard size is 16mm^2, so this is the one to use.

Note: Owing to the voltage drop, it may be necessary to make adjustments to the AVR such that a higher voltage is generated than that required at the point of use. However, if you are generating at 240V, a 12V drop to 228V—as in the above example—will not cause problems with any electrical appliances.

Costs summary

Cable costs are proportional to the amount of copper contained within the cable. When you are trying to minimize voltage-drop, take care to balance this economy against the increased costs associated with the higher grade cable that may be required. Also remember to give consideration to whether the cable is to be laid or strung. With larger schemes it is often cheaper to go for overhead lines—burial requires extra insulation, not to mention the labour and plant costs. However, bear in mind that buried cable is less susceptible to potentially expensive damage from adverse weather conditions.

SUMMARY

By this stage, you should have a fairly clear understanding of some of the main design and engineering aspects involved in the development of typical micro-hydro schemes. If you are going to build your own scheme, there are two further practical issues that need to be addressed: will it be legal, and how can you pay for it? These are discussed in the following two chapters.

Chapter 7
Legal and Environmental Aspects

INTRODUCTION

Hydro power clearly benefits the global environment, in terms of the reduction in greenhouse gas emissions and acid rain pollution. However, it can have a negative impact on the local environment. Hydro power involves the use of valuable water resources, and the needs of a hydro installation will often have to be weighed against other uses for the water. Additionally, the nature of the resource means that hydro schemes will often be situated in environmentally sensitive areas.

In most countries there are laws in place designed to protect the local countryside from potentially environmentally destructive developments. This chapter provides an outline of the various environmental and legal aspects to be considered in the development of micro-hydro schemes and summarizes methods for ameliorating any local environmental impacts. By considering your local environment early in the process, you should be able to ensure a smooth passage through the legalities surrounding the installation of any new scheme.

ENVIRONMENTAL IMPACT ASSESSMENT

An assessment of the likely environmental impacts of the development is one of the first steps to be taken. Regard should be given to the passage of fish; flood defence; effects on water quality; land drainage; residual flow in the river; and the effects on users upstream, in between the intake and outflow, and downstream. The scope of the assessment will depend on the scale of the development—larger schemes may have to take additional account of the impact on flora, fauna, noise levels, the landscape, social conditions, land use, and cultural heritage. Small-scale developments are far easier to manage environmentally. The area of watercourse disturbed is far smaller, and with proper consideration, the ecological and environmental effects can be negligible.

PLANNING PERMISSION

As with all developments, planning permission is generally required for a new hydro-electric scheme. An exception may be where there is no 'change of use', such as in the case of the refurbishment of an existing scheme.

The procedure is generally fairly straightforward. The first stage is to get in touch with your local planning office. They will require that you provide them with the details of your proposal, which may be a simple case of filling out a form and providing layout drawings. They will also advise which consultees will need to be approached and the scope of the environmental study. A small fee will be levied, the value depending on the land area used.

LICENCES

Some or all of the following licences may be required, depending on where the scheme is to be developed. In England and Wales these should be obtained through the Environment Agency, in Scotland through the Scottish Environmental Protection Agency, and in Northern Ireland through the Environmental Heritage Services. In other countries you should be able to find out through your local planning office.

Abstraction licence

Since water is such a vital resource, its use is generally restricted and subject to licensing control. The diversion of water from a river or stream by pipe or leat is termed abstraction. In England and Wales, any scheme abstracting in excess of 20m^3 per day will have to be licenced by the Environment Agency. Only the tiniest of schemes (20m^3/day = 0.23 l/s) are exempt; at the other end of the spectrum, large-scale hydro (5MW+) generators are levied an annually renewable volume-related charge. All schemes in between, whether they be a 200kW NFFO-assisted commercial operation or a 2kW stand-alone system, will incur similar costs and a fixed one-off fee in order to comply. In Scotland and Northern Ireland micro-hydro schemes do not require an abstraction licence yet.

Conditions will be imposed to protect the stream's ecology and the rights of other users. The minimum requirement will be to leave the 95 percentile flow (the flow that is exceeded in the river for 95% of the year) in the stream at all times; in some cases, the scheme may be restricted to taking 25% of available water above the 95 percentile flow. To minimise impact (and so licence conditions), it can be worth restricting the design flow of a scheme to less than the 50 percentile flow. As environmental awareness increases, control of water use is becoming tighter.

In England and Wales, the basic cost for abstraction is at least £250, which should cover the processing fee and public notices in the local press. Further costs may be incurred if flow monitoring is required. It is strongly advised that an informal agreement is made with the relevant agency before the formal application is placed.

Works in river permission
A works in river consent may be required for the intake works. Once details of the intake design are provided, consent is given based on conditions relating to care of the aquatic environment during construction. In the UK, the basic cost is a £50 processing fee.

Impounding licence
An impounding licence may be required if the scheme involves the construction of a new dam or weir of significant volume. Additional Land Drainage Consent may be required if the weir or dam is to be constructed in a tributary that feeds into a main river.

OTHER CONSENTS
Depending on the nature and location of the scheme, several additional consents may be required. This is likely to be established during consultation with the local planning authority.

Examples include:
* Building Regulation approval (with larger schemes)
* Special consents for construction on common land or near ancient monuments
* National Park Authority permissions
* Discharge Consent (Scotland)

ENVIRONMENTAL MITIGATION MEASURES
To ensure environmental compliance, some or all of the following measures may need to be adopted.

Provision of compensation flow
Most watercourses support fish, so provision must be made for their survival. There must always be a certain minimum flow of water (called the compensation flow) left in the stream between the intake and outlet. This is done by letting a percentage of the stream flow bypass the weir. Generally, the 95 percentile flow (Q95) is a minimum requirement for compensation flow.

Screening
Fish must be protected from the scheme, since any fish passing through the system may be killed. There is also a risk of causing water hammer damage to the penstock due to blocked jets. A typical maximum spacing between bars is 10mm, and the maximum approach velocity to the screen is 0.25m/s (when partially blocked), to stop fish getting through or caught on the screen. For low head, high flow schemes, acoustic fish scarers are sometimes used to keep fish away from inlets.

Fish passes
In larger weirs, it may be necessary to include a fish ladder to allow easy movement of fish past the weir. This is fairly easy to arrange with a run-of-the-river scheme, but more ingenious solutions are required for schemes with storage reservoirs. Some schemes include fish lifts, water locks, or fed fish passes. In some cases, acoustic fish scarers may be used to keep fish away from the intake.

On the above issues, you will generally be steered to an appropriate solution by the authority giving the right of use of the water.

Chapter 8
Economics

INTRODUCTION

Having looked at all the other aspects of establishing a micro-hydro power scheme, we now turn to the economics. The focus is on household systems, although where appropriate, consideration is also given to larger commercially-operated micro-hydro schemes.

OVERVIEW

Where a house is already connected to a mains electricity supply, the odds are that investing in a small micro-hydro scheme is not generally a good short-term financial investment. However, this will depend very much on the energy demand profile of the house and how that demand is currently met: is the scheme displacing gas-fired water and space heating, or electrically powered water and space heating, etc.? In the former case it would almost certainly not make economic sense, whereas in the latter it might. In the long term, however, and especially with a very good site, the installation of a micro-hydro scheme should be a good investment, whatever the case. Additionally, in situations where the instantaneous demand can be higher than the supply, the use of an ELC can allow the mains to cut in once the demand outstrips the scheme's supply, as described in chapter 6. Where supply outstrips demand, it may, in the near future, become feasible to sell any surplus to the Grid (see chapter 10).

Where there is no existing mains electricity supply, the installation of a good hydro scheme will almost certainly make better economic sense than investment in a Grid-connection. Grid-connections can be horrendously expensive, diesel generators have high running costs, and other renewables may not be appropriate. In economic terms, a stand-alone micro-hydro scheme is best seen as an addition to the house: an asset that increases the re-sale value of the property. A well designed scheme will have a lifetime of at least twenty years before major re-financing is required, so seen in this way, in times of often uncertain fuel prices, a scheme on a productive site should prove to be an excellent long-term investment.

When considering household schemes, a major factor determining viability will be how effectively the supply is utilised by that household's demands. Ideally, demand would match the supply, and there would be no need to balance the system. However, in all probability the demand will vary from several kilowatts to none, depending on the time of the day and month of the year. Of the balancing that is required, much can be achieved through the use of an ELC. However, to achieve the most economical balance, some effort will need to be made to manage the household loads, as discussed in chapter 3. This will distribute the demand more evenly throughout the day, minimising the amount of supply surplus.

To decide whether the scheme will be economically viable, a close estimate of the overall cost needs to be calculated, for which two main factors must be considered: capital costs and operation and maintenance costs.

Capital cost
The initial cost of the scheme will be the biggest hurdle for most people, as the entire capital outlay needs to be borne at the beginning of the project. Unlike other forms of renewable energy system, such as solar panels, it is not cost-effective to install hardware with a power rating much less than that which the house requires, with the intention of adding on larger equipment at a later date. There is an 'economy of scale' at play, whereby a scheme rated at 5kW, for example, may only cost 50% more than a scheme producing 2kW.

Operation and maintenance costs

Operation and maintenance (O&M) costs should also be considered, in addition to possible downtime due to flooding, blockage, or equipment failure. The likelihood of unplanned downtime is far less for well planned schemes, but it is still wise to have some spare bearings and belts handy. As a rule, bearings will need to be replaced once every four to seven years, and they should be greased every six months. Allow around £20–£50 per kW capacity, per year, to cover expenses and inconveniences during downtime.

SCHEME COSTS BREAKDOWN

Some of the typical costs for the major components have been outlined in chapters 5 and 6. Others, which are less easy to quantify, include those for the design and installation of a scheme. These vary enormously, depending on the site layout. There is variation in the distribution of the costs depending on the scheme's topography and head. As a rule, a high head site will require a less expensive turbine (either a Pelton, Turgo, or compact crossflow), but have greater penstock costs due to the need for increased length and pressure-rating. A low head scheme will use a more expensive turbine involving more complicated civil works, but with a shorter, lower pressure rated (and hence cheaper) penstock. On balance, higher head schemes tend to be more cost effective.

Cost breakdown examples are shown below, but these should only be considered to be general cases.

Domestic scheme

The following table is based on a site using a multi-jet Pelton turbine to produce 5kW on a 25m head site to supply the demand of a remote farmhouse.

Design	15%
Installation	12%
Civils hardware: concrete, powerhouse, building materials	15%
Penstock	20%
Turbine	20%

Alternator	10%
Electrical controls, cable & wiring	8%

The total cost of such a system would be around £25,000 with the system professionally installed, less if installed on a DIY basis.

Large commercial scheme
These cost distributions are based on a 200kW site using 4 x 50kW propeller turbines installed in the mid '90s.

As can be seen, the design costs are more or less fixed so they make up a smaller percentage of the total compared to the 5kW site. Because of the hardware expense of Grid connection, the electrical distribution costs increase. On the other hand, because of an 'economy of scale,' the electromechanical expenses drop slightly.

Design	10%
Installation	15%
Civils hardware: concrete, powerhouse, building materials	12%
Penstock	18%
Turbine	22%
Alternator	7%
Electrical controls, cable & wiring	15%

The total cost was around £300,000 with the scheme installed professionally.

PROFESSIONAL INSTALLATION
For Grid-connected and other larger schemes, some recourse to professional assistance is essential. However, for smaller schemes you may be able do the whole thing yourself. The degree of professional help required will depend on both your time and skills, and on the nature of the scheme. This section looks at the possibility of installing a small household-sized scheme by going for the 'turnkey' option—paying a specialist to complete the entire project.

Having the scheme designed and installed by a professional would certainly cost more than the DIY approach. However, there are a number of advantages, which, over the long-term, could offset the extra cost.

The most important consideration is that a professionally installed scheme should be of a guaranteed high quality, ensuring that it is an asset to the house. Seen in this way, the scheme is a quality addition to the property, similar in principal to professionally installed double glazing or solar water heating; such additions increase the overall value of the house. A well designed scheme should run happily for a minimum of twenty years before requiring a relatively modest and inexpensive overhaul. When the property is put on the market, the scheme will be an attractive asset rather than a high-maintenance liability.

The second major advantage is that the designer will be able to size the scheme to extract the maximum power at minimum equipment cost, avoiding the pitfalls open to the DIYer. The hidden pitfalls of the initially cheaper option can range from leaky pipes and inefficiently sized equipment, to a total washout in a flood due to poorly planned civil works.

It is important to note that the above points focus on the quality of the design and installation of the scheme. While the professional installer will have the advantage of perhaps many years of experience, there is no reason why dedicated DIYers, with plenty of time on their hands, should not be able to install a perfectly adequate scheme by themselves.

If the turnkey route is to be taken, it is worth noting that most micro-hydro scheme installers offer a free, brief, desk based site assessment based on the site's head and estimated year round flows. Before contacting a supplier it may be helpful to take several stream flow measurements, including, if possible, drought and flood flows. Also, supplying a large-scale map containing the proposed scheme's stream and catchment will help improve the accuracy of the assessment. Always ask a professional installer for a client reference list and proof of professional indemnity insurance.

GENERAL COST COMPARISONS WITH OTHER TECHNOLOGIES

Micro-hydro generally offers the cheapest long-term supply of electricity compared to the other forms of remote power supply. Photovoltaics are very expensive in terms of set up costs, while diesel systems have similar capital costs but are more than three times as expensive per unit of electricity produced.

The indicative data in Table 8.1 has been compiled by calculating the total capital cost involved to deliver a unit of electricity, taking into account the different lifespans of each source.

Table 8.1

Energy source	Price per unit (kWh)
Small-scale remote power supplies	
Micro-hydro	5p
Wind (inc batteries)	15p
Diesel	20p
Solar photovoltaic (inc batteries)	60p
Commercial supplies	
Large hydro	3p
Natural gas	3p
Landfill gas	3.5p
Large wind	3.5p
Coal	4p
Nuclear	5p+

ECONOMIC VIABILITY

Having assessed the likely costs of a scheme, the next stage will be to calculate whether or not it is worth going ahead and developing it. This section gives an idea of the returns which can be expected, using simple methods.

The scope of the economic assessment will depend very much on the nature and size of the project. For people with some spare cash, wishing to install a 1kW scheme in their garden, profitability may not be the first concern, and the quick assessment given here will be more than adequate. On the other hand, the developers of a £0.25m, 100kW scheme selling to the Grid, would be hard-pressed to attract the funding required unless they could provide a comprehensive financial evaluation. For such schemes, a fuller financial evaluation than what follows will be necessary.

Payback period

The most elementary financial indicator is the simple payback period. This gives the amount of time required for a project to have recouped its costs, either through electricity sales for Grid-connected schemes, or through electricity savings for autonomous schemes.

Simple payback period = $\dfrac{\text{capital costs (years)}}{\text{annual sales/savings - annual costs}}$

Example

A 2kW scheme is to be installed at a cost of £6,000. It is estimated that £750 worth of electricity will be generated annually and that annual operation and maintenance costs will amount to no more than £100.

Simple payback period   = $\dfrac{£6,000}{£750 - £100}$

=   9.2 years

This method is very simplistic. It takes no account of interest charges or the time-value of money—the fact that the original investment of £6,000 is worth more in the present than in the future.

Interest rates

In the real world, loans attract interest. To calculate the annual repayments, R, required for a capital sum, C, at a certain interest rate, ir, to be paid off in n years, the following sum is used:

Annual repayments   =   C   x $\dfrac{(ir) \times (1 + ir)^n}{(1 + ir)^n - 1}$

Example

Using the data from the simple payback period example and applying an interest rate of 8%, the necessary annual payment will be as follows:

Annual repayments $\qquad$ = $\qquad$ $\dfrac{£6,000 \times 0.08 \times (1 + 0.08)^{9.2}}{(1 + 0.08)^{9.2} - 1}$

$\qquad$ = $\qquad$ £946 per year

Over 9.2 years this will amount to £8,703, hence the payback period will be:

$\qquad$ $\dfrac{£8,703}{£750 - £100}$ $\qquad$ = $\qquad$ 13.4 years.

PROJECT FINANCING

If you don't have all the required capital, then you will need to raise a loan. A micro-hydro scheme can pay itself off in less than ten years and is a good long term investment, but first you will have to convince a lender, probably your bank manager, of this fact. The bank manager's primary interest is the security of the bank's money; before arranging a meeting, it might pay to collate your property deed records, for instance, to show that the investment will be secured by the net value of your property.

You will need to provide an accurate costing based on the quotes of suppliers or installers to demonstrate that the proposed loan will be adequate. A variety of repayment options will be provided; it is worthwhile giving detailed consideration to each. The better value-for-money the loan, the shorter the scheme's payback period. Shopping around is recommended.

As a rule of thumb, the annual loan repayments should be more or less equal to the annual sum that, without the scheme, would have been spent on mains electricity consumption.

Grants

There may be special grants/ loans available, such as the Clear Skies, Community Renewable Investment and Farm Improvement Loan in the UK. These are worth looking into seriously before embarking on a project. However the funding is usually only partial, the application procedures long-winded, and the required criteria often tight. Installers will be able to offer advice on this and other schemes.

Financing Grid-connected commercial schemes

Grid-connected schemes can be commercially viable, but money can be lost as easily as it is made. Several opportunities exist, from selling power to an electricity supply company, to supplying a single large 24-hour consumer or a number of smaller consumers at peak times.

If a scheme requires a partially unsecured loan, the criteria for loaning money are much stricter and you will need to provide a comprehensive financial evaluation to have a chance of securing the necessary funding. The production of such a document is an involved procedure which is often best left to an experienced professional. However, to reduce costs, the following tasks could be undertaken:

• talk to electricity companies / brokers to gauge interest and likely price per unit paid including ROCs for short and long term contracts

• discuss Grid connection with local grid company

• compile quotes from reputable suppliers and installers

There are special investor funds that attract people who understand the extra risks involved in backing hydro schemes. A reputable installer should be able to provide further information.

For further details on Grid-connected hydro schemes, see 'Commercial micro-hydro schemes' in chapter 10.

SUMMARY

Having analysed the economics of the scheme, the next stage is either to abandon the idea as a pipe dream, or... take the plunge and go ahead and build.

Chapter 9
Commissioning and Operation
& Maintenance (O&M)

INTRODUCTION

Having built your micro-hydro scheme, you will no doubt be eager to get it going. This chapter focuses on small domestic schemes using an ELC; it is split into two parts. The first looks at the commissioning stage of the scheme and details the steps necessary to ensure its successful implementation. The second outlines the key factors to be considered when devising a sensible operation and maintenance schedule. Brief tips on safety are included throughout. These tips are not comprehensive; a full risk assessment should be carried out for each operation and the authors take no responsibility for actions taken by readers. Be careful!

COMMISSIONING

Having installed all the components, the final stage is to get the scheme up and running. While it may be tempting to complete the commissioning as quickly as possible, try to be patient. Time is necessary to allow for adequate performance testing, and you will need to document the original running conditions for future reference. Commissioning has two stages: pre-startup and startup.

Pre-startup

Consider each element of the scheme in turn: intake, channel, penstock, valves, ballast load, alternator, electrical connections, turbine.

Intake

Ensure the intake is clear of any loose debris or stones that may pass through or block the screen during the initial startup. Ensure that any de-silting tanks are clean.

Safety precautions: The intake is potentially lethal to inquisitive youngsters—screens and covers must be fixed securely so as to be impossible for children to remove.

Penstock checks

Check that the civil works for all the anchor blocks are complete and cured and pipe joints are secure.

It is highly likely that some debris or small wildlife will have found its way into the penstock during its construction, even if preventative measures have been taken. To flush it out, run a reasonable quantity of water down its length for three to four hours before the penstock is connected to the turbine, being careful not to flood the powerhouse in the process!

To fully commission the penstock a pressure test should be carried out as follows:

Close the main inlet valve and turbine valves, and slowly start filling the penstock. Using the pressure gauge inside the power-house, stop filling at regular intervals and check for leaks or movement. It's helpful to have a few spotters up the line to alert the pipeline filler in the event of any problems. Once the penstock appears to be full, wait one minute to ensure that any air trapped during filling has a chance to escape.

To check the penstock for leaks, walk the entire length, checking at each joint (where left unburied) for signs of seepage. This is particularly important with underground pipes, since it is obviously far more difficult to locate and repair a leak once the pipe has been buried. If a leak is found, even a minor one, it is best to empty the penstock, attend to the leak, and repeat the penstock filling procedure. Standard plastic sleeves and couplers are available if the joint cannot be sealed.

At this stage the penstock will be operating under its static head, but it is desirable to increase the pressure by a factor of 50% (at least 10–20%) in order to simulate the water hammer that would occur during a minor jet blockage or a rapid valve closure. This can be achieved through either one of the two following methods:

• Take a commercially available bung with a connection for joining to a length of hose, and fit it into the mouth of the intake. If the hose is carefully filled while the penstock is full, the overall head on the scheme will increase and will correspond to the water level in the elevated length of hose, which needs to be filled to the required test head.

• Use a blank flange on the pipe at the intake and close the MIV. Use a water pump with over pressure by-pass to pressurise the pipe slowly by filling with additional water. Consider the forces due to the increase in pressure and make sure all anchor blocks are designed to resist this force.

Safety precautions: Medium to high-head penstocks are potentially dangerous due to the high pressures involved; if they break, they break violently. Ensure that the number of people near the lower penstock length is kept to a minimum, and that they have been informed of the potential danger. Stay clear of any second-hand components that have been used in the penstock until they have been fully tested.

Ballast load

For household supply schemes with an ELC, the ballast load usually consists of a space and/or water heater, located in the house, which consumes the dumped power from the controller. Check that they are securely wired and any thermostats are bypassed. Most importantly, to prevent any fire risk, ensure that any space heaters are not covered. If there is a chance of space heaters being turned off accidentally while the plant is being run, remove the plug and wire the cable into the supply permanently. Grid-connected schemes will have a different dump configuration which needs to be checked for wiring integrity.

Alternator

Check that the rotation of the alternator is smooth, and ensure that its ventilation port is not obstructed. This is best done by allowing enough water to flow through the system to rotate the alternator to about 5–10% of its rated speed, after pressure testing the pipe (see later).

If the alternator is run whilst the inner windings are very damp, there is a danger of it burning out. It is best to dry it out thoroughly before commissioning. As a rule, keep the system running as much as possible once it is commissioned, even if just to produce a few amps. In addition to providing minimal power for lighting, this will keep the alternator dry.

Electrical connections

Check that all connections are tight and sufficiently insulated, and that the polarities are correct.

Turbine

Ensure all manifold couplers, nuts, and bolts are tightened correctly and re-check wiring for integrity and safety. Check that all moving parts are free from obstruction and that the bearings have been adequately lubricated.

If a belt drive is used, then re-check the belt tension.

Safety tips: Ensure that all moving parts are covered by guards so that it is impossible to touch them accidentally. Fit a lock to the powerhouse door to deter inquisitive youngsters.

Penstock pressure testing

Startup

Once all the pre-commissioning checks have been completed, slowly open the turbine valve and gradually bring the alternator up to speed (with some generators you may have to set the AVR first). Use a minimum of water to achieve this—just enough to cause the ELC to cut in (the load controller comes into operation at a pre-set alternator speed). You can usually hear when this has happened. At this stage, check that the voltage is stable and that it is within the desired range. Some of the possible causes of voltage variation are mechanical speed fluctuations, a slipping belt drive, or a faulty or incorrectly set AVR. Sometimes the ELC may require adjustment.

Continue to slowly open one valve fully, watching the pressure gauge to check that water hammer (oscillation on the pressure gauge) is limited to less than 10% of static head.

Check the bearings for excess heat. This will occur if they are misaligned or over-greased.

Once the scheme is running, record all the readings and settings. Typical headings could include the following:

- static pressure gauge reading (bars or kPA)
- running pressure gauge reading, noting any variation (bars or kPA)
- leakages
- runner speed (rpm)
- alternator frequency (Hz)
- alternator speed (rpm)
- alternator current (amps), voltage, and power (at full and part flows)
- current, amps and power at load
- mechanical or electrical trip operation

Note that during this initial running period the chance of a jet blockage is quite high, so be aware and double check the power output values after the other checks are complete.

Carry this whole procedure out for each jet or power level until full power is achieved.

OPERATION & MAINTENANCE

While hydro schemes generally require very little maintenance in order to avoid untimely breakdowns or expensive damage, it is crucial to plan and carry out what is necessary.

This section covers the basic requirements for effective O&M based on a household sized scheme; it concludes with a suggested maintenance schedule.

Log book

It is recommended that you keep a weekly record of activities carried out, with comments on system operation. This is will be useful for diagnostic purposes, should a fault occur. The suppliers of the scheme's components should recommend appropriate O&M schedules in the operating manuals accompanying their equipment.

Intake and settling tank
Unless a maintenance free screen is used without a settling tank, the intake screens and settling tanks will require cleaning whenever they begin to impair the functioning of the scheme. Generally, this is a simple matter of raking screens and opening the gate or valve at the base of the settling tank, digging and flushing away any debris that has accumulated.

Maintenance work may well be required on intakes that are located in flood-prone streams. In some cases, repairs may be needed; more usually, it will be a simple matter of clearing debris and ensuring the watercourse has not become obstructed.

Jet and screen blockages
If the output of the turbine falls below that expected, there are a handful of likely causes:
- the turbine is running out of water due to a blocked screen
- there is a general lack of water (turbine flow too high for the available river flow indicated by a pressure gauge drop)
- a jet has become blocked (in which case the pressure gauge remains steady)

If you suspect that the turbine is running out of water, reduce the flow and observe the pressure gauge. If there is a shortage, the gauge will show an increase in pressure as the pipe begins to refill. Check whether or not the intake has become obstructed, especially during autumn—chances are that a clean is all that will be required. A blocked jet is more of a problem, as many turbines have limited jet access. Time-consuming though it may be, it will always be possible to unblock the jet, and some turbines do have relatively easy access, often via an inspection hole.

Safety precautions: Find out from the turbine suppliers how best to proceed, and always follow their safety instructions. Always empty the penstock of water if there is any chance of its restraining supports being weakened/dismantled during the operation.

Penstocks

Steel penstocks require periodic painting to avoid external rusting, while exposed plastic needs to be protected from direct sunlight by providing cover. Penstock supports need to be regularly checked for integrity.

Depending on the stream, often there is a gradual build up of slime on the inside of the pipe which can dramatically increase head loss, reducing system performance. This can be cleaned off by flushing a foam swab or 'pig' down the pipe.

Buried pvc or polyethylene penstocks generally require no additional maintenance unless they begin to leak.

Turbines

Turbines require very little maintenance (other than bearings, if fitted), as long as the water supply is kept clean. However, if things do get into the turbine, it will be necessary to stop the equipment and remove them.

Francis and crossflow turbines are particularly prone to blade wear from silt particles in the water. Where a settling tank has been used, it is important to keep the water clean by ensuring that the tank is emptied regularly. Both of these turbines have relatively fragile blades which are prone to damage from debris, so ensure that the trash racks are well designed.

Problems with freezing

During the winter months, there is the risk of the water freezing. If the pipeline freezes solid whilst full of water, considerable damage to the penstock and manifold may result. Freezing is unlikely to occur if the turbine is run continuously, because the water temperature will be maintained by friction losses as it flows through the pipe. Freezing may well occur if the flow stops. Therefore it is better to keep the machine running; if it does need to be stopped, then drain the system of water completely by blocking the inflow of water and leaving the valves fully open. Alternatively, it may be possible to allow a gentle trickle of water—sufficient to prevent freezing—to flow through.

Belts and bearings
Check bearings every month by listening to the turbine in operation. To listen to a bearing, use a long screwdriver resting on the bearing housing with the handle pressed against the ear to amplify its internal noise. When the bearings are getting worn, the bearing noise will change from the normal smooth hum to a gravely rumble. The turbine should not be run at all if the bearings are worn, as internal damage to the alternator may result.

The bearings should be designed for at least five years of maintenance-free operation. However this may be reduced by excessive condensation and water ingress. Hence the operator should be able to discern between healthy and worn bearings. Regarding the belts, for good performance they should be:
- clean and free of oil and grease
- tightened to the correct tension
- correctly aligned

Valve seals
If leaking occurs around the stem, then the shaft seal should be replaced. However, this is not critical, so it can be left to coincide with other maintenance. The seal is a simple lip seal. Always use grease during re-assembly.

Electrical connections
Every six months check all connections are tight.

Tools and spare parts
The advantages of having spare bearings and belts are obvious. Other equally useful necessary parts and tools include the following:
- waterproof grease
- belts
- alternator and turbine bearings
- fuses for ELC
- AVR spares
- fuses
- a section of spare penstock
- penstock paint (if applicable)

- multi-grade oil
- grease-gun
- bearing removal tool
- spanners, screwdrivers, hacksaw, etc.
- rake for trash rack

Maintenance schedule

An example of a maintenance schedule is outlined below. Note that it does not include the Grid-connect hardware for a Grid-connected scheme.

Item	Normal conditions	Floods
Weir & intake		
Check and clean screens bar	daily	daily
Check for boulder damage	monthly	daily
Check for leaks, undercutting		
Settling tank (where used)		
Maintain sluicing device	monthly	
Drain and clean	monthly	daily
Channel (where used)		
Inspect for leaks and overflowing	weekly	daily
Drain and clean	quarterly	
Penstock		
Check for leaks	biannually	
Repaint	2 yearly	
Check for corrosion	2 yearly	
Check for slime build up	annually	
Powerhouse		
Check for noise or vibration	monthly	
Check bearing operation & grease	biannually	
Check belt tension	quarterly	
Check for manifold or valve leakage	monthly	

Item	Normal conditions	Floods
Check alternator ventilation	biannually	
Inspect changeover switches	biannually	
Replace belts and bearings	every 5 years	

SUMMARY

With a bit of luck the scheme should now run, with occasional simple maintenance, for at least twenty trouble-free years.

Chapter 10
Further Applications

INTRODUCTION

Micro-hydro power is not simply about using orthodox and proven methods of producing small quantities of electricity for your own use. This chapter looks at further applications of micro-hydro power.

The first two sections expand upon two possibilities already considered:

- commercial hydro schemes—selling your power
- water-wheels—a traditional and aesthetically pleasing way of producing power.

Next the chapter examines the esoteric world of heat pumps, and how a hydro-scheme can be adapted to provide a highly efficient form of heating. This is followed by 'pumps-as-turbines' and how to produce hydro-electricity, very simply, by running a standard electrical pump in reverse. Finally, the hydraulic ram pump is discussed. A simple, extremely reliable, well established technology, the hydraulic ram pump harnesses the power contained in a flow of water to raise a portion of that water to an elevated level. It is a water-powered water pump.

COMMERCIAL MICRO-HYDRO SCHEMES

Because they generate relatively small amounts of electricity, micro-hydro schemes are not usually feasible as commercial ventures. However, there are exceptions, and this section looks at the options available to developers who are considering selling some or all of their electricity.

There are two basic options for selling the electricity that is generated: through the national electricity transmission system (the 'Grid') to an electricity supply company or user; or directly to a third-party using either your own transmission infrastructure or theirs.

Selling through the Grid

In the UK and across much of Europe, the electricity industry is currently being deregulated and de-nationalised into three separate areas of private business: generation (Generators), transmission (National Grid Company and Distribution Network Operators) and supply (Electricity Supply Companies). Most electricity is generated by very large (up to 4000MW) power stations, which supply electricity directly to the very high voltage (400,000V) National Grid. From the National Grid, the electricity is stepped down to lower voltages for distribution in local networks. The local networks are owned and operated by a number of Distribution Network Operators (DNOs). Since privatisation in England and Wales in 1990 (slightly later in Scotland), the industry has opened up to allow the supply of far smaller quantities of electricity directly to the lower voltage local distribution networks. This is partly as a result of an increase in renewable electricity generation. It is at this level that a Grid connected micro-hydro scheme would supply its electricity—directly into an 11kV or 33kV local network.

Hardware

Setting up the hardware required for sale to the Grid is an expensive business and, as a rule, not considered to be economically viable for schemes of 25kW or less. Grid codes exist to regulate the connection of generation to the Grid. G59 is generally applied to micro hydro above 12kW, whilst there are less demanding G77 (single phase up to 5kW) and G82 (3 phase up to 16A per phase) codes for smaller schemes. For a start, the electricity sold will have to be synchronous with that already in the Grid, in addition to this, adequate and approved electrical protection equipment is required. Don't expect to see much change from £8000 for the Grid connection equipment. This cost may drop as protection rules are changed for very small generators and/or mass produced (e.g. solar) interface units can be adapted.

Selling to an Electricity Supply Company

To sell to an electricity supply company you have to negotiate a power purchase agreement (PPA). Except for old Non Fossil Fuel Obligation (NFFO) or Scottish Renewables Obligation (SRO) contracts the value of a PPA is generally made up of three separate income components that renewable electricity can generate: the base electricity unit (BEU), the Renewable Obligation Certificate (ROC) and Levy Exemption Certificate (LEC).

Base Electricity Unit. The electricity market is very competitive and current average BEU price is around 2p per kWh. However, this will be open to negotiation, and if, for instance, the scheme is very remote, feeding a weak grid and/or can be configured such that it is only generating at times when the local demand is high (during business hours, winter time, etc.), then a better price may be secured.

Renewables Obligation Certificates

The Renewables Obligation and Renewables Obligation (Scotland) support mechanisms launched in April 2002 in the UK are having a big effect on the electricity industry. All electricity supply companies are obliged by law to supply a proportion of their electricity from renewable sources, increasing to 15% by 2015, (with an aspiration of 50% by 2050!), compared to the approximately 3% supplied in 2003. Renewables Obligation Certificates (ROCs) with a guaranteed minimum value of 3p are assigned to each kWh of eligible renewable electricity. Where a supply company does not supply its quota of renewable energy it has to pay a buy-out fee of 3p per kWh shortfall into a fund, which is then paid out pro-rata to those holders of ROCs. Since there is currently a shortfall in the renewables market ROCs are currently worth more than the 3p minimum. Importantly for micro hydro, renewable electricity is eligible for ROCs whether it is sold through the Grid or not.

Levy Exemption Certificates.
The Climate Change Levy is imposed on electricity from fossil fuels at 0.43 p/unit and since renewables are exempt a portion of this saving (usually about 50%) can be attributed to renewable electricity (adapted from Scottish Renewables Forum Financing Renewable Projects Briefing note).

Advantages of this situation are that there is a guaranteed sizeable market (at least in the short term) for renewable electricty and that current rates paid for renewable electricity are generally higher than NFFO contracts (particularly for wind power). Some disadvantages are that the market is new, complex, continually changing and there are no guaranteed long term secure contracts. To finance a new project it will usually be necessary to negotiate a minimum 5 year contract. Although the total value may be over 6p in the short term, the best reported deals for 5 year contracts are 4 to 5p per unit.

The NFFO and SRO schemes
The NFFO (Non Fossil Fuel Obligation) and SRO (Scottish Renewables Order) schemes were put in place by the UK Government in 1990 to safeguard the future of nuclear power generation, after privatisation, by obliging the Regional Electricity Companies (RECs) to purchase electricity from non-fossil fuels at a premium price. As renewable electricity is also non-fossil, this too was included in the scheme. There were five renewables NFFO orders, in 1990, 1991, 1994, 1997 and 1998. In each, potential renewable developers had to submit a bid price—the lowest price at which they are prepared to sell their electricity. The schemes offering the best prices were then accepted, and the local RECs obliged to buy the electricity at that indexed linked fixed rate for a 15 year (except for the first round) period. Some schemes were not built and the contracts are transferable. Unused hydro NFFO contract prices range from 5p to 7p per unit, which is probably better than you'll get in a ROC based PPA.

Net metering
Net metering is a possible option for sites with on-site load, where

Grid connect case study: 200kW low-head system

Based at a farm on the site of an old 35kW system, this scheme is a low head system built into the existing weir. The operating head is 6m, whilst the total flow is 6m³/s. Four propeller turbines and two generators are used to convert the water power, operating independently to match the available flow. The combined output is 200kW, and this power is fed into the local electricity network. A Grid connection and flow controller automatically operate the scheme without the need for permanent staffing.

The total cost of the scheme was around £240,000 installed. The net annual income through sale of electricity is in the order of £30,000, and the scheme was developed on the basis of a contract received under the NFFO. The simple payback of around eight years illustrates the viable nature of Grid-connected schemes. The major components of the scheme will last in excess of thirty years.

the scheme supplements electricity supplied from an existing Grid-connection. This is the situation with most Grid connected solar systems. The concept is quite new to the UK, but has been around in several states of the US since the early 1980s. The local network operator (DNO) is obliged to allow connection of small generation as long as the appropriate Grid codes are followed. Basically, net metering is an import/export arrangement with an electricity company, whereby small generators of electricity can sell on any surplus not used on site, and buy back from the Grid at times of shortfall. In effect, the Grid is used in the same way as a battery-store. An import/export meter is installed for recording units exported and imported. You'll need to set up a deal with an electricity supply company (some of the 'green' ones are keen) and will probably receive about two thirds of the price you pay per unit imported for the exported ones.

Green electricity trading pools

As of September 1998, all UK electricity consumers are technically free to choose from where they buy their electricity. As a result of ever-increasing environmental awareness, growing numbers of these consumers demand, and are prepared to pay a premium to ensure, that their electricity comes from renewable resources. This has lead to the establishment of green electricity supply companies, organisations that will act as broker between the producers and users of renewable energy. Sale through such an organisation can offer the best returns.

Selling direct to a third party
One possible exception is selling the electricity, via the Grid, directly to a third party. The third party may be a nearby business that runs industrial machinery, for instance. As a result of deregulation, the DNOs are now obliged to rent their lines to small generators, thereby allowing for the sale of electricity in this manner. The chief advantage of such a set-up is that there is no 'middleman'. The difficulties of identifying customers, the costs for connection to the Grid, for use of the system, metering, administration and probably having to become a licensed supply company should not be overlooked.

Selling outside the National Grid
It is possible to sell direct to consumers, without recourse to the Grid. This option will only be possible if they are geographically close to the scheme, but has the advantage that it avoids all the costs associated with the use of the Grid, and the ROCs can still be sold separately. On the downside you may have to become a registered supply company, there is a fair administrative burden and the customer will need a back up system.

SUMMARY
In the recent past, only large-scale commercial electricity generation existed in the UK; however, deregulation of the electricity supply industry has meant that far smaller amounts of electricity can be generated on a commercial basis. This does not mean to say that the process is simple or even likely to be profitable, just that some of the legal instruments preventing it have been, or are in the process of being, removed. Whilst facilitating a boost in renewable generation nationally, the ROCs support mechanism is generally barely sufficient to allow financing of new professional-build micro hydro, but is an added bonus to small schemes.

WATER-WHEELS

Water-wheels have been discussed earlier, albeit briefly, as the forerunners to the modern turbine, both in the introductory chapter, and in chapter 5. While it is perfectly feasible to use water-wheels for electricity generation, they have not, so far, been described in much detail. A typical micro-hydro electric scheme extracts the power from a relatively low flow and fairly high head of water, using a compact, high-speed turbine, supported by a fairly standard civil works infrastructure. It has little in common with a scheme using the water-wheel as the prime mover. Typical water-wheels exhibit the following characteristics:

- the head is very low, being limited by the wheel's diameter
- the flow is usually fairly high to compensate for the limited head
- water-wheels tend to spin very slowly: to generate electricity they must be coupled up to expensive very high gearing mechanisms
- water-wheels are generally quite large, material- and labour-intensive, and are not mass-produced. As a consequence they tend to be expensive

As a general rule, when compared with a modern low-head turbine, the installation of a new water-wheel for the purpose of electricity generation does not make financial sense. However, there are exceptions: the Pedley Wood Conservation Trust, an environmental charity based in Cheshire, have recently developed an efficient and low-cost electricity generating water wheel—the Pedley Wheel. Designed for use in less developed countries, the first unit was installed in Sri Lanka in early 1998. At a total cost of £6000, the wheel provides up to 2.5kW of mains quality power under a head of 4m. Many of the design complications involved with traditional water-wheels have been overcome: the Pedley Wheel is the result of over seven years' research and development.

With more traditional water-wheels, financial viability may not be the only consideration—if aesthetic sensibilities are taken into account, the picture may look quite different. Indeed, with many of the water-wheels running in the UK, the production of electricity is very much a secondary priority, quite over-shadowed by the ambience their presence evokes.

─────────────── The Pedley Wheel ───────────────

The low cost electricity-generating water wheel (Bill Smith)

The remainder of this section describes some of the possibilities for making use of the power generated by traditional water-wheels, and will be of most relevance to those thinking of refurbishing an existing mill.

Overcoming the technical problems
The inherent problem with water-wheels with respect to electricity generation is that they are low-speed, high torque devices (since power = speed x torque). The 'buckets' of a typical overshot wheel move at 2–3m/s, while the wheel may have a diameter of around 3–5m. This translates to a rotational speed of between 7–20rpm. Since commonly available generators need to be driven at 1500rpm (or at the very least 750rpm), a speed increasing drive of at least 75:1 is usually required. Acquiring such a drive, particularly an efficient one, is not cheap. Indeed, if an overshot water wheel and a speed increasing drive are costed at new prices, the combination is typically two to five times the price of an equivalent modern low-head turbine.

—The Rock—

Fig. 10.1

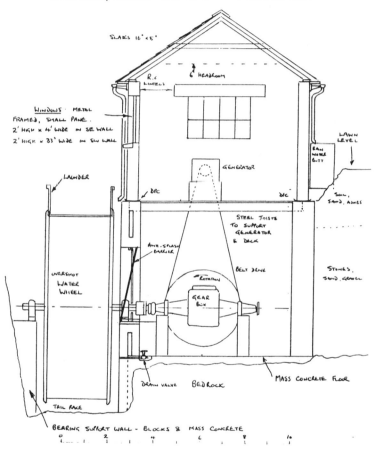

POWER HOUSE — SECTIONAL ELEVATION LOOKING S W

SCALE 1" TO 2'

CONSTRUCTION — GEAR ROOM OR BASEMENT: MASS CONCRETE FLOOR ON BEDROCK. CONCRETE BLOCK WALLS; NW & NE SOLID WITH 2" WATERPROOF CEMENT MORTAR DPC; SW WALL 6" BLOCKS; SE WALL 4" BLOCKS WITH 2" CAVITY. ACCESSIBLE WALLS INSIDE & OUT RENDERED. GENERATOR ROOM OR GROUND FLOOR 6" BLOCK WALLS, RENDERED EXTERNALLY & GABLES. ROOF ON CONVENTIONAL TIMBER RAFTERS &c. PLASTICS RW GUTTERING & DOWN-PIPES. METAL WINDOW FRAMES, SMALL PANE. DOOR HALF OR PART GLAZED, SMALL PANES.

SLATES 16" × 8"

R.C LINTELS

6 HEADROOM

WINDOWS: METAL FRAMED, SMALL PANE.
2' HIGH × 4' WIDE IN SE WALL
2' HIGH × 33' WIDE IN SW WALL

LAWN LEVEL

RAIN WATER BUTT

LAUNDER

GENERATOR

DPC

DPC

SOIL, SAND, ASHES

STEEL JOISTS TO SUPPORT GENERATOR & DECK

ANTI-SPLASH BARRIER

BELT DRIVE

STONES, SAND, GRAVEL

OVERSHOT WATER WHEEL

ROTATION

GEAR BOX

DRAIN VALVE

BEDROCK

MASS CONCRETE FLOOR

TAIL RACE

BEARING SUPPORT WALL — BLOCKS & MASS CONCRETE

0   2   4   6   8   10

Sectional elevation of water-wheel powered hydro-electric scheme at 'The Rock', Devon (Courtesy of Commader G. Chapman)

However, in most cases the wheel is not bought new; old water-wheels can be easily refurbished. Alternatively it would be quite feasible to put one together from scratch—in comparison with water turbines, water-wheels are fairly uncomplicated to make.

Drive technicalities
The most popular DIY option is to use the main differential from a tractor (it is quite possible to get hold of a second-hand tractor gearbox fairly inexpensively)—which gives an 8:1 ratio—combined with a belt drive second stage. In effect, one stub axle is locked, whilst the other is driven by the water-wheel and the power is taken from the high-speed shaft. To improve efficiency, oil pump-fed direct lubrication can be fitted, since energy is lost in sump oil churning. In addition to using a tractor gearbox, there are several other ways of achieving the speed increase required: a straight six stage belt drive; a rim gear (a small pinion that runs on a circular rack mounted on the rim of the wheel) with single stage belt drive; and a gearbox and integral gearbox-motor as generator.

Remember to consider the low-speed, high-torque nature of water-wheels—don't rely purely on power ratings of gearboxes. Also consider the thrust forces relating to helical gears (the sideways force due to gears having angled teeth), particularly if running gearboxes in the reverse direction to that for which they were designed.

HEAT PUMPS
If your main energy requirement is in the form of heat, you may wish to consider installing a hydro-powered heat pump. The heat pump is a device that will deliver more heat energy than the electrical or mechanical energy used to drive it; each kWh of electricity used in driving the pump results in several kWh of heat output.

How does it work?
The principle behind a heat pump's operation rests in the fact that all matter, providing it is above a certain absolute minimum temperature, contains energy in the form of heat. The heat pump uses the energy contained in a large, low temperature heat source,

to increase the temperature of a smaller heat sink. By reducing tho temperature of, say, a 1000m³ volume of air by 1°C, it is possible to increase the temperature of a smaller, 100m³ volume of air by 10°C. This, thought of in reverse, is exactly the same mechanism by which a fridge, freezer, or air-conditioning unit works. A fridge simply transfers the heat energy contained in the food and air inside (the heat source) to the kitchen outside (the heat sink).

How is the heat energy transferred?
The heat is transferred by a circulating fluid known as a 'refrigerant'. Refrigerants have a very low boiling point and, as with any liquid, this point varies according to the pressure it is under. When a liquid boils and becomes vapour, it absorbs heat; when it reverts back to a liquid it releases that heat. In the heat pump, the liquid refrigerant reaches its boiling point and becomes vapour as it passes through the heat source. The vapour is then pressurised by the compressor; this increases its boiling point sufficiently for the vapour to return to a liquid state as it passes through the condenser. On returning to its original liquid state, the refrigerant releases its heat.

——————————Heat Pump Circuit——————————

Fig. 10.2

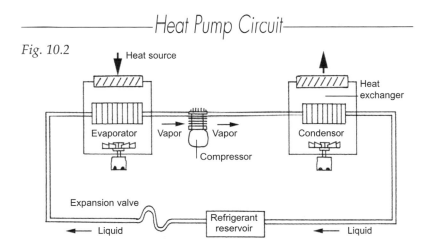

Energy is required to drive the pump, but, as explained earlier, this is less than the heat energy the pump can provide. The relationship between the heat energy output and the electrical or mechanical energy input is referred to as the coefficient of performance or COP. Typical COPs of small to medium scale heat pumps are 3-5—a heat pump that consumes 3kW of power for moving the heat will supply 9–15kW of heat. The less the temperature difference between the heat source and heat demand, the better the COP: thus underfloor heating is good to use with a heat pump as the floor, in effect, becomes a large low temperature radiator. However, heat pumps powered by Grid electricity have little advantage over gas, as the inefficiencies at the power station usually equal the COP.

Hydro-powered heat pumps
Heat pumps can work well with hydro power schemes—the first ever heat pump, made in 1926, was hydro-powered. The water can be used both to power the scheme and to provide an abundant heat source for the heat pump. It is quite possible to drive the pump mechanically and thereby do away completely with electricity generation and the associated energy losses. Indeed, if the purpose of the hydro scheme is simply to produce heat, then going for a mechanically-driven heat pump will provide several times more heat energy than generating electricity to power bar-heaters. In such instances, the use of a water-wheel will be just as effective as the use of a turbine for producing the motive power required to drive the compressor.

One final point: most heat pumps can, at the flick of a switch, be run in reverse. The device that provides warmth in the winter can also keep you cool in the summer!

Although they are an exciting idea, water-powered heat pumps are not generally readily available. Specialist designers and suppliers may have to be contacted, or you might wish to experiment and make one yourself.

PUMPS-AS-TURBINES
An electric pump, run in reverse, can be used as a turbine. All-in-one units, which are mass-produced, have an integral pump that

can serve as a turbine, and a motor that doubles as a generator. Thus there is wide scope for selecting a unit that is well matched to the head and flow conditions at any given site. The pumps are likely to be relatively inexpensive, easy to install, and have readily available spare parts.

However, not all pumps are suitable, as can be seen in Table 10.1.

TABLE 10.1: Suitability of different pump types

Centrifugal pumps	End-suction	most suitable
	In-line	less efficient
	Double suction	less efficient
	With round casing	inefficient
Self-priming pumps		suitable if valve removed
Submersible pumps	Dry-motor, jacket cooled	suitable
	Dry-motor, fin cooled	unsuitable
	Wet motor, borehole type	unsuitable
Positive displacement pumps	(e.g. gear pumps, mono pumps, piston pumps)	unsuitable

Adapted from A Williams, *Pumps-as-turbines* (1995 IT Publications).

Increasingly, many of the major pump manufacturers should be able to provide data on the suitability of their various models. The major limitation of using pumps-as-turbines is that, for a given head, a fixed flow rate is required at all times; if this cannot be ensured, then generation will not be efficient. They are most appropriate for medium or low head sites, as a possible alternative to a multi-jet Pelton or a crossflow turbine.

HYDRAULIC RAM PUMPS

While the hydraulic ram pump cannot be construed as micro-hydropower in the traditional sense (it does not produce electricity), it is a valid technique for making use of the power in water.

The hydraulic ram, or hydram, is a device that uses the power available from a flowing stream to pump a small proportion of the water up a considerable height above the stream. In short, it is a water-powered water pump. It is designed to take water from a high-flow, low-head source and divert some of that flow to a high head. The hydram works by taking water in a small pipe from a point up-stream that is 1–20m higher than the hydram, so that the water arrives at the ram under pressure.

A brief history

The hydraulic ram was invented in 1793 by Montgolfier, of hot-air balloon fame. Manufacture in Britain began in 1820, and they soon became extremely common as a means of pumping drinking water supplies. Most were installed to service isolated rural homes and farms, though occasionally whole villages depended on them.

The arrival of mains water supplies, or cheap conventional pumps and cheap power to run them, nearly led to the demise of the hydraulic ram in the UK. But now, as our faith in mains water is coming increasingly into question and we are trying to reduce our energy demand, the ram is set for a revival.

How does it work?

Water is taken from a point upstream that is higher than the ram, down the drive pipe (see figure 10.3). This height difference is called the supply head, and means that the water arriving at the ram is under pressure. If the waste valve is open, this pressure pushes water through the valve and back into the stream. However, the water flowing through tends to drag the valve with it, until the valve slams shut.

With this exit cut off, there is a sudden increase of pressure in the valve box (water hammer); the pressure opens the delivery valve and pushes water up into the air vessel. When the pulse of pressure passes, the delivery valve shuts, preventing the water

Hydraulic Ram Pump

Fig. 10.3

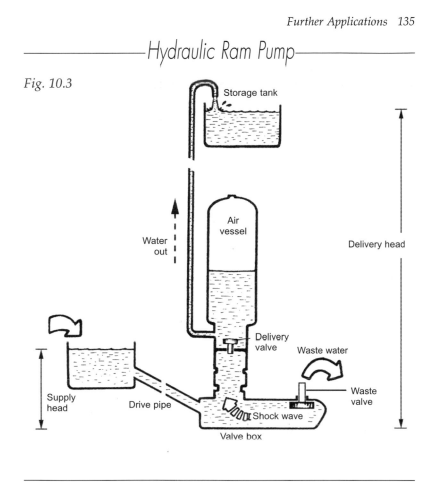

from running back. The waste valve then falls open, allowing the cycle to start again. With each cycle, the pressure in the air chamber builds up until the water is pushed up the delivery pipe to wherever it is needed.

Application

Any site that is close to running water, but uphill from that supply, may well be able to benefit from the ram. The supply head might typically be 1–6m, but it could be less than 0.5m. With a 6m supply head, the delivery head could be as much as 60m.

A commercial hydraulic ram, properly installed, protected, and maintained, may provide a water supply for over a hundred years

136 Going with the Flow

without replacement. It needs very little maintenance and is obviously suitable for irrigation purposes, and ideal for use where any suitable source of water exists. It will continue to pump fresh water as long as there is water to pump, without emitting pollutants or greenhouse gases, and without running up your fuel bills. (Adapted from the CAT *Hydraulic ram pump* tipsheet).

Glossary of Terms

alternating current (ac): Electric current in which the direction of flow reverses continually, very quickly. Mains electricity is ac, and in the UK it has a frequency of 50Hz (hertz)—this means that it changes through one forward and one backward cycle 50 times per second.

alternator: An electromechanical device for generating ac electricity.

altimeter: A device used to measure altitude, useful when calculating large heads of water.

ampere (amp, A): The unit of electrical current measurement, the flow rate of electrons in a circuit.

annual hydrograph: A graphical representation of flow patterns for a given watercourse over a year.

asynchronous generator: A generator that produces electricity of a variable frequency.

automatic voltage regulator (AVR): A device that keeps the voltage of a generator constant.

average daily flow (ADF): The average flow rate in a given watercourse calculated on the basis of measurements taken over a year. Measured in m^3/s or l/s.

ballast load: A device, usually an electric space heater or water heater, to which generated power in excess of that required directly for other needs is diverted.

breastshot water-wheel: A water-wheel fed by water entering at the level of the wheel's axle.

calibrated pressure gauge: A device used for measuring pressure, useful for calculating head.

catchment area: The area of land above the intake that provides the springs and rainwater run-off from which water is collected and passes into the stream.

channel: An open canal sometimes used to convey the water over the distance from the intake to the settling tank at the entry to the penstock.

coefficient of performance (COP): The relationship between heat energy output and electrical or mechanical energy input; usually considered with reference to heat pumps.

current (electrical): The amount of electricity flowing through a circuit, measured in amps or amperes (A).

Distribution Network Operator (DNO): A company which operates part of the electricity distribution grid or network.

design flow: The flow rate used as the basis on which a micro-hydro scheme is designed.

direct current (dc): Electric current that flows in one direction only through a circuit. Batteries provide dc electricity.

diversion weir: A structure built into a watercourse to divert a portion of the water from the flow and through a hydro scheme.

draught tube: A conical tube attached the underside of reaction turbines designed to create a suction force which has the effect of increasing the net head.

dumpy-level: An optical device used in conjunction with a measuring staff to establish differences in height between points.

electrical power: The rate of delivery or consumption of electrical energy at any instant. The electrical power (P) of a device is the product of the voltage (V) across it and the current (I) passing through it (P = IV). Power is measured in watts (W); 1000W = 1 kilowatt (kW), 1000kW = 1 Megawatt (MW), 1000MW = 1 Gigawatt (GW).

electrical energy: The amount of electrical work done over a given period. The product of the average power (P) and the length of time (T) the power is produced or consumed for (E = PT) and is measured in watt-hours (Wh). 1000Wh = 1 kilowatt-hour (kWh) = 1 unit of mains electricity. 1000kWh = 1 megawatt-hour (MWh).

electronic load controller (ELC): A device that keeps an alternator's speed constant by providing a constant electrical load on the generator.

flow duration curve: A graphical representation of the range of flow rate experienced by a given watercourse throughout a year.

flow rate: The rate at which a body of water is flowing, measured in either cubic meters per second, or litres per second.

frequency: Cycles per second (see alternating current), measured in hertz (Hz).

generator: An electromechanical device that converts motive power into electrical power.

gigawatt: See electrical power.

head: The vertical distance between the intake and the turbine, measured in metres.

head loss: The difference between the real head and the effective head caused by frictional energy losses occurring in the transmission of the water through the penstock.

head race canal: An alternative term to describe the channel.

hertz (Hz): See alternating current.

impulse turbine: A turbine in which a jet (or jets) of water impinges on the blades or buckets of its runner to cause rotation.

induction generator: An asynchronous generator.

intake: The point in a hydro scheme where the water is diverted from the stream or river; often built into a diversion weir.

intake orifice: The orifice built into the intake and sized to allow the design flow to pass into the channel, but restrict the flow during floods.

isohyetal map: A map of an area showing rainfall level contour lines.

kilowatt: See electrical power.

Levvy Exemption Certificate: Certificate giving exemption from the Climate Change Levvy – a fixed rate 'carbon tax'.

megawatt: See electrical power.

Non Fossil Fuel Obligation (NFFO):): A 90's scheme under which UK regional electricity companies are obliged to purchase non fossil fuels at a premium price.

overshot water-wheel: A water-wheel fed by water flowing onto it from above.

penstock: The pipe used in a hydro scheme through which the water is conveyed, under the pressure of the head, to the turbine.

pool selling price (PSP): The commercial rate at which electricity is sold from the pooled supply of the UK's major electricity generators. The PSP varies on a half-hourly basis according to national levels of supply and demand.

powerhouse: The building in a hydro scheme housing the turbine and electricity-generating equipment .

power purchase agreement (PPA): An agreement with an electricity supplier for the purchase of electricty

rainfall map: Another word for an isohyetal map.

reaction turbine: A water turbine where the runner is spun as a result of a pressure drop created across its blades.

regional electricity company (REC): A company that owns and operates a local electricity network (now DNO).

Renewables Obligation: UK support mechanism for renewable electricity.

runner: the 'wheel' of a turbine through which the power of the water is converted into rotational shaft power.

settling tank: A large tank often used at the end of a channel to slow the water down sufficiently to allow debris to settle out prior to the water's entry into the penstock.

skimmer: A pole or similar object secured across the surface of the water just before the intake to prevent the ingress of floating debris

spillway: A section of the channel designed to allow a controlled overflow of water. In the event of flooding or a channel blockage, a spillway should prevent the channel from bursting its banks.

stoplog: A gate used to prevent the flow of water through the intake.

surge pressure: The pressure that occurs when the water pressure in the penstock exceeds that that it would normally be due to head. It occurs when the flow of water is suddenly interrupted.

synchronous generator: A generator that produces electricity of a constant frequency.

tailrace: The pipe or channel to return water from the powerhouse back to the watercourse.

theodolite: A device used in the mapping of areas to provide accurate height and position measurements.

transmission cable: Buried or overhead electric cable used for transmitting the electricity from the generator to the load.

trash rack: A screen, fitted across the entry to the penstock, that prevents leaves, debris and fish from entering the system and damaging the turbine.

voltage: A measure of the electrical potential across any two points in a circuit, such as the terminals of a battery or generator; the force that makes the electrical current flow around the circuit.

watercourse: The body of water—stream or river—from which the power for a scheme is to be harnessed.

water intake diversion weir: A construction in the watercourse that seals the riverbed and stabilizes the flow, allowing some of it to be diverted into the scheme.

water mill: A water-wheel with grinding apparatus attached to its shaft for milling flour, etc.

water turbine: A device for converting the power in a body of water into high speed rotational motive power.

water-wheel: A device that converts the power in a body of water into slow speed, high torque, rotational motive power.

wing walls: The walls that make up the side(s) of the intake, parallel to the water flow.

Conversion Factors

Length: millimeters (mm), metres (m), kilometres (km)
1 inch = 25.4mm = 0.0254m
1 foot = 305mm = 0.305m
1 mile = 1609m = 1.609km

Area: square millimetres (mm^2), centimetres (cm^2), and metres (m^2); hectares
1 inch^2 = 645mm^2=0.645 x $10^{-3}m^2$
1 acre = 4047m^2
1 hectare = 10,000m^2

Volume: cubic millimetres (mm^3), centimetres (cm^3) and metres (m^3); litres (1)
1 inch^3 = 16.4 x 10^3mm^3 = 16.4 x $10^{-6}m^3$
1 pint (UK) = 0.568 1= 568ml
1 cm^3 = 1000mm^3
1m^3 = 1000 1

Mass: grams (g), kilograms (kg), tonnes (t)
1 pound (lb) = 0.454 kg
1 short ton = 907.2 kg
1 long ton = 1016 kg
1 tonne = 1000 kg

Velocity: metres per second (m/s), kilometres per hour (km/h)
1 foot per second = 0.3048 m/s
1 mile per hour = 0.447m/s = 1.61 km/h

Volume flow rate: cubic metres per second (m^3/s)
1 cubic foot per second = $0.283m^3/s$
1 gallon (UK) per second = $0.00455m^3/s$

Pressure: pascals (Pa), newtons per square metre (N/m^2), bars
1 pascal = $1 \ N/m^2$
1 pound per square inch (psi) = 6895 Pa
1 bar = $10^5 \ Pa$
1 atmosphere = 14.7 psi = 1.01325×10^5 Pa = 1.01325 bar

Work and energy: joules (J), kilojoules (kJ), megajoules (MJ)
1 foot pound = 0.113 J
1 calorie = 4.186 J
1Btu = 1.055 kJ
1kWhr = 3.6 MJ

Power: watts (W), kilowatts (kW), megawatts (MW)
1 foot pound per second = 1.356W
1 horsepower = 746W = 0.746kW

Index